Almost periodic functions and differential equations

Almost periodic functions and differential equations

B.M.LEVITAN & V.V.ZHIKOV

Translated by L.W.Longdon

CAMBRIDGE UNIVERSITY PRESS
Cambridge
London New York New Rochelle
Melbourne Sydney

Published by the Press Syndicate of the University of Cambridge,
The Pitt Building, Trumpington Street, Cambridge CB2 1RP
32 East 57th Street, New York, NY 10022, USA
296 Beaconsfield Parade, Middle Park, Melbourne 3206, Australia.

Originally published in Russian as *Pochti-periodicheskie funktsii i differentsial'nye uravneniya* by the Moscow University Publishing House 1978
Assessed by E. D. Solomentsev and V. A. Sadovnichii
First published in English, with permission of the Editorial Board of the Moscow University Publishing House, by Cambridge University Press 1982

Printed in Great Britain at the University Press, Cambridge

Library of Congress catalogue card number: 83 4352

British Library Cataloguing in Publication Data
Levitan, B.M.
Almost periodic functions and differential equations.
1. Periodic functions
I. Title II. Zhikov, V.V. III. Pochtiperiodicheskie funktsii i differentsial'nye uravneniya. *English*
515.8 QA331
ISBN 0 521 24407 2

Contents

Preface

The theory of almost periodic functions was mainly created and published during 1924–1926 by the Danish mathematician Harald Bohr. Bohr's work was preceded by the important investigations of P. Bohl and E. Esclangon. Subsequently, during the 1920s and 1930s, Bohr's theory was substantially developed by S. Bochner, H. Weyl, A. Besicovitch, J. Favard, J. von Neumann, V. V. Stepanov, N. N. Bogolyubov, and others. In particular, the theory of almost periodic functions gave a strong impetus to the development of harmonic analysis on groups (almost periodic functions, Fourier series and integrals on groups). In 1933 Bochner published an important article devoted to the extension of the theory of almost periodic functions to vector-valued (abstract) functions with values in a Banach space.

In recent years the theory of almost periodic equations has been developed in connection with problems of differential equations, stability theory, dynamical systems, and so on. The circle of applications of the theory has been appreciably extended, and includes not only ordinary differential equations and classical dynamical systems, but wide classes of partial differential equations and equations in Banach spaces. In this process an important role has been played by the investigations of L. Amerio and his school, which are directed at extending certain classical results of Favard, Bochner, von Neumann and S. L. Sobolev to differential equations in Banach spaces.

We survey briefly the contents of our book. In the first three chapters we present the general properties of almost periodic functions, including the fundamental approximation theorem. From the very beginning we consider functions with values in a metric or Banach space, but do not single out the case of a finite-dimensional Banach space and, in particular, the case of the usual numerical almost periodic functions. Of the known proofs of the approximation theorem we present just one: a proof based on an idea of Bogolyubov. However, it should be noted that another instructive proof due to Weyl and based on the theory of compact operators in a Hilbert space appears in many textbooks on functional analysis.

Chapter 4 is devoted to the theory of N-almost periodic functions. In comparison with the corresponding chapter of the book *Almost-Periodic Functions* by B. M. Levitan (Gostekhizdat, Moscow (1953)), we have added a proof of the fundamental lemma of Bogolyubov about the structure of a relatively dense set.

Chapter 5 is concerned with the theory of weakly almost periodic functions developed mainly by Amerio.

Chapter 6 contains, as well as traditionally fundamental questions (the theorem of Bohl–Bohr about the integral, and Favard's theorem about the integral), more refined ones, for instance, the theorem of M. I. Kadets about the integral.

We mention especially Chapter 7 whose title is Stability in the sense of Lyapunov and almost periodicity. The two chapters that follow it are formally based on it. Actually, we use only the simplest results, and when there is a need to refer to more difficult propositions we give independent proofs. Therefore, Chapters 6–11 can be read independently of one another.

Chapter 8 contains Favard theory, by which we mean the theory of almost periodic solutions of linear equations in a Banach space. In Chapter 9 the results from the theory of monotonic operators are applied to the problem of the almost periodicity of solutions of functional equations. In Chapter 10 we give another approach to the problem of almost periodicity. Finally, Chapter 11 is slightly outside the framework of the main theme of our book. In it we give one of the possible abstract versions of the classical averaging principle of Bogolyubov.

Chapters 1–5 were written mainly by B. M. Levitan, and Chapters 6–11 by V. V. Zhikov.

The authors thank K. V. Valikov for his assistance with the reading of the typescript.

Translator's note

This translation has been approved by Professor Zhikov, to whom I am grateful for correcting my mistranslations and some misprints in the original Russian version.

Professor Zhikov has asked me to mention that the theory of Besicovitch almost periodic functions is not reflected fully enough in the book, since this theory has recently been applied in spectral theory and in the theory of homogenisation of partial differential equations with almost periodic coefficients. The additional references are, in the main, concerned with this theme.

L. W. Longdon

1 Almost periodic functions in metric spaces

1 Definition and elementary properties of almost periodic functions

Throughout the book J denotes the real line, X a complete metric space, and $\rho = \rho(x_1, x_2)$ a metric on X.

Let $f(t): J \to X$ be a continuous function with values in X; we denote the range of f, that is, the set $\{x \in X: x = f(t), t \in J\}$, by $\mathscr{R}_f$, and its closure by $\bar{\mathscr{R}}_f$.

Definition 1. A set $E \subset J$ of real numbers is called *relatively dense* if there exists a number $l > 0$ such that any interval $(\alpha, \alpha + l) \subset J$ of length l contains at least one number from E.

Definition 2. A number τ is called an ε-*almost period* of $f: J \to X$ if

$$\sup_{t \in J} \rho(f(t+\tau), f(t)) \leq \varepsilon. \tag{1}$$

Definition 3. A continuous function $f: J \to X$ is called *almost periodic* if it has a relatively dense set of ε-almost periods for each $\varepsilon > 0$, that is, if there is a number $l = l(\varepsilon) > 0$ such that each interval $(\alpha, \alpha + l) \subset J$ contains at least one number $\tau = \tau_\varepsilon$ satisfying (1).

Every periodic function is also almost periodic. For if f is periodic of period T, then all numbers of the form nT $(n = \pm 1, \pm 2, \ldots)$ are also periods of f, and so they are almost periods of f for any $\varepsilon > 0$. Finally, the set of numbers nT is relatively dense. It is easy to produce examples of almost periodic functions that are not periodic, for instance, $f(t) = \cos t + \cos t\sqrt{2}$.

We prove some of the simplest properties of almost periodic functions; these are straight-forward consequences of the definition.

Property 1. *An almost periodic function $f: J \to X$ is compact in the sense that the set $\bar{\mathcal{R}}_f$ is compact.*
Proof. It is sufficient to prove that for any $\varepsilon > 0$, $\mathcal{R}_f$ contains a finite ε-net for $\mathcal{R}_f$. Let $l = l(\varepsilon)$ be the length in Definition 3 corresponding to a given ε. We set

$$\mathcal{R}_{f;l} = \{x \in \mathcal{R}_f\colon x = f(t), -l/2 \leqslant t \leqslant l/2\}.$$

From the continuity of f it follows that the set $\mathcal{R}_{f;l}$ is compact; we show that it is an ε-net for the set $\mathcal{R}_f$. Let $t_0 \in J$ be chosen arbitrarily, and take an ε-almost period $\tau = \tau_\varepsilon$ such that $-l/2 \leqslant t_0 + \tau \leqslant l/2$, that is,

$$-t_0 - l/2 \leqslant \tau \leqslant -t_0 + l/2.$$

Then

$$\rho(f(t_0 + \tau), f(t_0)) \leqslant \varepsilon.$$

Because $t_0 + \tau \in [-l/2, l/2]$, the set $\mathcal{R}_{f;l}$ is an ε-net for $\mathcal{R}_f$, as we required to prove.
Remark. For numerical almost periodic functions (that is, when $X = R^1$) and for almost periodic functions with values in a finite-dimensional Banach space, Property 1 reduces to the following: if f is an almost periodic function, then $\mathcal{R}_f$ is bounded.

Property 2. *Let $f: J \to X$ be a continuous almost periodic function. Then f is uniformly continuous on J.*
Proof. We take an arbitrary $\varepsilon > 0$ and set $\varepsilon_1 = \varepsilon/3$ and $l = l(\varepsilon_1)$. The function f is uniformly continuous in the closed interval $[-1, 1+l]$, that is, there is a positive number $\delta = \delta(\varepsilon_1)$ (without loss of generality we may assume that $\delta < 1$) such that

$$\rho(f(s''), f(s')) < \varepsilon_1 \tag{2}$$

whenever $|s'' - s'| < \delta$, $s', s'' \in J$. Now let t', t'' be any numbers from J for which $|t' - t''| < \delta$. We take a $\tau = \tau_{\varepsilon_1}$ with $0 \leqslant t' + \tau_{\varepsilon_1} \leqslant l$, that is, $-t' \leqslant \tau_{\varepsilon_1} \leqslant -t' + l$. Then $t'' + \tau_{\varepsilon_1} \in [-1, 1+l]$. We set $s' = t' + \tau_{\varepsilon_1}$ and $s'' = t'' + \tau_{\varepsilon_1}$. From (1), (2) and the triangle inequality we have

$$\rho(f(t''), f(t')) \leqslant \rho(f(t''), f(s'')) + \rho(f(s''), f(s')) + \rho(f(s'), f(t')) < \varepsilon.$$

Property 3. *Let $f_n : J \to X$, $n = 0, 1, 2, \ldots$, be a sequence of continuous almost periodic functions that converges uniformly on J to a function f. Then f is almost periodic.*

Proof. We take an arbitrary $\varepsilon > 0$ and let $n = n_\varepsilon$ be such that

$$\sup_{t \in J} \rho(f(t), f_{n_\varepsilon}(t)) \leqslant \varepsilon/3. \tag{3}$$

Let $\tau = \tau[f_{n_\varepsilon}]$ denote an $(\varepsilon/3)$-almost period of the function f_{n_ε}. Then it follows from (1), (3), and the triangle inequality that

$$\begin{aligned} \rho(f(t+\tau), f(t)) &\leqslant \rho(f(t+\tau), f_{n_\varepsilon}(t+\tau)) \\ &\quad + \rho(f_{n_\varepsilon}(t+\tau), f_{n_\varepsilon}(t)) + \rho(f_{n_\varepsilon}(t), f(t)) \\ &\leqslant \varepsilon \end{aligned}$$

for all $t \in J$. This proves that f is almost periodic because the set of almost periods $\tau[f_{n_\varepsilon}]$ is relatively dense.

Property 4. *Let $x = f(t)$ be a continuous almost periodic function with values in a metric space X, and $y = g(x)$ be continuous on $\bar{\mathcal{R}}_f$ with values in a metric space X_1. Then $g[f(t)]$ is an almost periodic function with values in X_1.*

Proof. Since the set $\bar{\mathcal{R}}_f$ is compact and the function $g(x)$ is continuous on $\bar{\mathcal{R}}_f$, $g(x)$ is uniformly continuous on $\bar{\mathcal{R}}_f$. Therefore, for all $\varepsilon > 0$ there exists a $\delta = \delta(\varepsilon) > 0$ such that for all $x', x'' \in \bar{\mathcal{R}}_f$ with $\rho(x', x'') \leqslant \delta$ we have

$$\rho_1(g(x''), g(x')) \leqslant \varepsilon.$$

Therefore, if τ is a δ-almost period for $f(t)$, then

$$\rho(f(t+\tau), f(t)) \leqslant \delta,$$

and so

$$\rho_1(g(f(t+\tau)), g(f(t))) \leqslant \varepsilon.$$

Corollary. *Let f be a continuous almost periodic function with values in a Banach space X. Then $\|f(t)\|^k$ is a continuous numerical almost periodic function for all $k > 0$.*

Property 5. *Suppose that f is an almost periodic function with values in a Banach space X. If the (strong) derivative f' exists and it is uniformly continuous on J, then f' is an almost periodic function.*

Proof. The proof uses the concept of an integral of a vector-valued function. In the case of continuous functions this is very simple because the Riemann integral exists with the usual fundamental properties (see, for example, G. E. Shilov, *Mathematical Analysis. Functions of a Single Variable*, Part 3, Ch. 12, § 12.5). By hypothesis, the derivative f' is uniformly continuous, and so for all $\varepsilon>0$ there is a $\delta=\delta(\varepsilon)>0$ such that $\|f'(t')-f'(t'')\|<\varepsilon$ whenever $|t'-t''|<\delta$. Therefore, if $1/n<\delta$, then

$$\left\|n\left[f\left(t+\frac{1}{n}\right)-f(t)\right]-f'(t)\right\|=\left\|n\int_0^{1/n}[f'(t+\eta)-f'(t)]\,\mathrm{d}\eta\right\|$$
$$\leqslant n\int_0^{1/n}\|f'(t+\eta)-f'(t)\|\,\mathrm{d}\eta<\varepsilon.$$

Consequently, the sequence of almost periodic functions $\phi_n(t)=n[f(t+1/n)-f(t)]$ converges uniformly on J to $f'(t)$. Now we only need to use Property 3.

2 Bochner's criterion

The main results of this section are also valid for almost periodic functions with values in an arbitrary metric space X. But for simplicity we shall assume that X is a Banach space. We shall use the following notation:

X denotes a complex Banach space; $x, y, z, \ldots$ are elements of X, and $\|x\|$ is the norm of $x\in X$. $C(X)$ denotes the Banach space of continuous bounded functions $f:J\to X$ with the norm

$$\|f(t)\|_{C(X)}=\sup_{t\in J}\|f(t)\|,$$

and $\mathring{C}(X)$ is the subspace of $C(X)$ consisting of almost periodic functions. Let us note that the spaces $C(X)$ and $\mathring{C}(X)$ are invariant under translations, that is, $C(X)$ ($\mathring{C}(X)$) contains together with $f=f(s)$ the function $f^t(s)=f(s+t)$ for all $t\in J$.

1. Bochner's theorem. *Let $f:J\to X$ be a continuous function. For f to be almost periodic it is necessary and sufficient that the family of functions $H=\{f^h\}=\{f(t+h)\}$, $-\infty<h<\infty$, is compact in $C(X)$.*

Proof. (*a*) *Necessity.* We assume that f is an almost periodic function (see § 1, Definition 3). We denote by $\{r\}$ the set of all rational points on J and let $\{f^{h_n}\}=\{f(t+h_n)\}$ be an arbitrary sequence of functions from H. By using Property 1 and applying the diagonal process, we can select from the sequence $\{f(t+h_n)\}$ a subsequence (we denote it again by $\{f(t+h_n)\}$) which converges for any $r\in\{r\}$. We prove that the sequence $\{f(t+h_n)\}$ converges in $C(X)$. We take an arbitrary $\varepsilon>0$ and let $l=l_\varepsilon$ be the corresponding length. Let

$\delta = \delta(\varepsilon)$ be chosen in accordance with Property 2. We subdivide the segment $[0, l]$ into p segments Δ_k $(k = 1, 2, \ldots, p)$ of length not greater than δ, and in each Δ_k we choose a rational point r_k. Suppose that $n = n_\varepsilon$ is chosen so that

$$\|f(r_k + h_n) - f(r_k + h_m)\| < \varepsilon \tag{4}$$

for $n, m \geqslant n_\varepsilon$ and $k = 1, 2, \ldots, p$. For every $t_0 \in J$ we find a $\tau = \tau_0$ such that

$$0 \leqslant t_0 + \tau \leqslant l \Leftrightarrow -t_0 \leqslant \tau \leqslant -t_0 + l.$$

Suppose that the number $t'_0 = t_0 + \tau$ falls in the interval Δ_{k_0} and that $r_{k_0} \in \Delta_{k_0}$ is the rational point chosen earlier. Then by our choice of δ we have

$$\begin{aligned} &\|f(t'_0 + h_n) - f(r_{k_0} + h_n)\| < \varepsilon, \\ &\|f(t'_0 + h_m) - f(r_{k_0} + h_m)\| < \varepsilon. \end{aligned} \tag{5}$$

It follows from (4) and (5) that

$$\begin{aligned} &\|f(t_0 + h_n) - f(t_0 + h_m)\| \\ &\quad \leqslant \|f(t_0 + h_n) - f(t'_0 + h_n)\| + \|f(t'_0 + h_n) - f(r_{k_0} + h_n)\| \\ &\qquad + \|f(r_{k_0} + h_n) - f(r_{k_0} + h_m)\| + \|f(r_{k_0} + h_m) - f(t'_0 + h_m)\| \\ &\qquad + \|f(t'_0 + h_m) - f(t_0 + h_m)\| < 5\varepsilon. \end{aligned}$$

Since $t_0 \in J$ was chosen arbitrarily, the last inequality implies that the sequence $\{f(t + h_n)\}$ converges in $C(X)$, that is, the set H is compact in $C(X)$.

(*b*) *Sufficiency.* We assume that the family $\{f(t + h)\}$, $-\infty < h < \infty$, is compact in $C(X)$ and prove that $f(t)$ is almost periodic (in the sense of Definition 3, § 1). First of all we show that f is a bounded function. For if this were not the case, then we could find a sequence of numbers h_n for which $\|f(h_n)\| \to \infty$. But then neither the sequence $\{f(t + h_n)\}$ nor any subsequence of it would be convergent at $t = 0$. From the boundedness of f it follows that the family of functions $\{f^h\} = \{f(t + h)\}$, $-\infty < h < \infty$ can be regarded as a set in $C(X)$.

By a criterion of Hausdorff, for all $\varepsilon > 0$ there are numbers $h_1, h_2, \ldots, h_p$ such that for all $h \in J$ there is a $k = k(h)$ such that

$$\sup_{t \in J} \|f(t + h) - f(t + h_k)\| < \varepsilon. \tag{6}$$

From (6) we have

$$\sup_{t \in J} \|f(t + h - h_k) - f(t)\| < \varepsilon,$$

that is, the numbers $h-h_k(h)$ $(k=1,2,\ldots,p)$ are ε-almost periods for $f(t)$. Now we only need to prove that the set of numbers $h-h_k$ is relatively dense. We set

$$L=\max_{1\leq k\leq p}|h_k|.$$

Then

$$h-L\leq h-h_k\leq h+L,$$

and since h is arbitrary this inequality implies that every interval of length $2L$ contains an ε-almost period for f.

2. Now we are going to deduce further properties of almost periodic functions that are obtained more simply from Bochner's criterion than from our definition.

Property 6. *The sum $f(t)+g(t)$ of two almost periodic functions is almost periodic. The product of an almost periodic function $f(t)$ and a numerical almost periodic function $\phi(t)$ is almost periodic.*

Proof. Let $\{h_n\}$ be an arbitrary sequence of real numbers. Firstly we extract from it a subsequence $\{h'_n\}$ such that the sequence of functions $\{f(t+h'_n)\}$ converges, and then a subsequence $\{h''_n\}$ of $\{h'_n\}$ for which the subsequence of functions $\{g(t+h''_n)\}$ is convergent. Then, clearly, the subsequence $\{f(t+h''_n)+g(t+h''_n)\}$ is convergent. Similarly, the product can be proved to be an almost periodic function.

Let $X_1, X_2, \ldots, X_n$ be Banach spaces, and let $X=\prod_k X_k$ be their cartesian product, that is, the Banach space with elements $x=(x_1, x_2, \ldots, x_n)$ and the norm

$$\|x\|=\sum_{k=1}^{n}\|x_k\|.$$

It follows easily from Bochner's criterion that if $f_1(t), f_2(t), \ldots, f_n(t)$ are almost periodic functions from J into $X_1, X_2, \ldots, X_n$, then the function $f(t)=(f_1(t), f_2(t), \ldots, f_n(t))$ is an almost periodic function from J into X. The next property is easily deduced from this remark.

Property 7. *Let $f_1(t), f_2(t), \ldots, f_n(t)$ be almost periodic functions from J into Banach spaces $X_1, X_2, \ldots, X_n$, respectively. Then for every $\varepsilon>0$, all the functions $f_1(t), f_2(t), \ldots, f_n(t)$ have a common relatively dense set of ε-almost periods.*

Proof. Suppose that τ is an ε-almost period for $f(t)=(f_1(t), f_2(t), \dots, f_n(t))$, that is,

$$\|f(t+\tau)-f(t)\|_X = \sum_{k=1}^{n} \|f_k(t+\tau)-f_k(t)\|_{X_k} < \varepsilon$$

for all $t \in J$. Obviously, for this τ we have

$$\|f_k(t+\tau)-f_k(t)\| < \varepsilon \quad (k=1,2,\dots,n),$$

as we required to prove.

3. The next property gives a condition for the compactness of a set of functions from $\mathring{C}(X)$, and is known as Lyusternik's theorem.

Lyusternik's theorem. *A set $M \subset \mathring{C}(X)$ is compact if and only if the following three conditions are satisfied:*

(1) *For every fixed $t_0 \in J$ the set*

$$E_{t_0} = \{x \in X : x = f(t_0), f \in M\} \subset X$$

is compact.

(2) *The set M is equicontinuous, that is, for every $\varepsilon > 0$ there is a $\delta = \delta(\varepsilon)$ such that $\|f(t') - f(t'')\| < \varepsilon$ whenever $|t' - t''| < \delta$ for all $f \in M$.*

(3) *The set M is equi-almost periodic, that is, for every $\varepsilon > 0$ there is an $l = l_\varepsilon$ such that every interval $(\alpha, \alpha + l) \subset J$ contains a common ε-almost period for all $f \in M$.*

Proof. (*a*) *Sufficiency.* The proof is exactly the same as that of the necessity for the conditions in Bochner's theorem.

(*b*) *Necessity.* By the criterion of Hausdorff, for every $\varepsilon > 0$ M contains a finite ε-net: $f_1, f_2, \dots, f_n$. Therefore, for all $f \in M$ there is a k_0, $1 \leqslant k_0 \leqslant n$, such that

$$\sup_{t \in J} \|f(t) - f_{k_0}(t)\| < \varepsilon. \tag{7}$$

For any $t_0 \in J$, from (7) we obtain

$$\|f(t_0) - f_{k_0}(t_0)\| < \varepsilon,$$

and so the finite set of elements $f_1(t_0), f_2(t_0), \dots, f_n(t_0)$ forms a finite ε-net for the set E_{t_0}. Consequently, E_{t_0} is compact in X, that is, condition (1) of Lyusternik's theorem holds. Condition (2) follows from the uniform continuity of each $f_k(t)$ $(k = 1, 2, \dots, n)$ on J and from (7). Finally, condition (3) follows from (7) and Property 7.

Remark. For numerical almost periodic functions, condition (1) of Lyusternik's theorem can be restated as follows: the set E_{t_0} is bounded.

3. The connection with stable dynamical systems

Suppose that we are given a 1-parameter group of homeomorphisms of a metric space X, $S(t): X \to X (t \in J)$. If for any $x \in X$ the corresponding trajectory $x^t = S(t)x$ is a continuous function $J \to X$ we shall call $S(t)$ a *dynamical system* or *flow*.

A flow $S(t)$ is called *two-sidedly stable* or *equicontinuous* if the transformations $S(t)$ $(t \in J)$ are equicontinuous on every compact set from X.

The next property is obtained from Bochner's criterion.

Property 8. *Every compact trajectory of a two-sidedly stable flow is an almost periodic function.*

Proof. We set $f(t) = S(t)x$. Since a trajectory is compact, we can extract from any sequence $\{f(t_n)\}$ a fundamental subsequence $\{f(t'_n)\}$. The transformations $S(t)$ are equicontinuous on the set $\bar{\mathcal{R}}_f$, and so

$$\sup_{t \in J} \rho(f(t+t'_m), f(t+t'_n)) \leq \varepsilon$$

whenever $\rho(f(t'_m), f(t'_n)) \leq \delta$, that is, Bochner's criterion holds.

The converse holds in a certain sense: with each almost periodic function $f: J \to X$ can be associated a compact trajectory of a two-sidedly stable dynamical system. For if we consider in $C(X)$ a system of translates, then the trajectory $f^t = f(s+t)$ is compact. Since the distance between two elements of $C(X)$ is invariant under a translation, we have an isometric and so two-sidedly stable flow. It is worth noting that the difference between isometry and two-sided stability is essentially insignificant; if a two-sidedly stable flow is defined on a compact space X, then it can be made isometric by choosing the following metric

$$d(x_1, x_2) = \sup_{t \in J} \rho(S(t)x_1, S(t)x_2).$$

It is easy to see that the metric d is invariant under translation and topologically equivalent to the original metric ρ.

Let $f: J \to X$ be an almost periodic function. We denote by $\mathcal{H} = \mathcal{H}(f)$ the closure of the trajectory $f^t = f(s+t)$ in $C(X)$, and are going to show that $\mathcal{H}$ is minimal in the sense that any trajectory is everywhere dense in it. Suppose that $\hat{f} = \hat{f}(s)$ is any element from $\mathcal{H}$. Then for some sequence $\{t_m\} \subset J$ we have

$$\sup_{s \in J} \rho(f(s+t_m), \hat{f}(s)) \leq 1/m.$$

Therefore,

$$\sup_{s \in J} \rho(f(s), \hat{f}(s - t_m)) \leq 1/m,$$

that is, $\hat{f}(s - t_m) \to f(s)$ uniformly with respect to $s \in J$. The closure of the trajectory $\hat{f}^t$ contains f, and so it coincides with $\mathcal{H}$.

4 Recurrence

The minimal property of an almost periodic function proved in the last section is in fact a very simple property of abstract trajectories.

1. Let X be a Hausdorff topological space.

We shall call a 1-parameter semigroup of continuous operators $S(t): X \to X$ $(t \geq 0)$ simply a semigroup, and shall use the symbols x^t, $x(t)$ to denote the semitrajectory $S(t)x$ $(x \in X, t \geq 0)$. A function $x(t)$ is called a *trajectory* of a semigroup $S(t)$ if $x(t+\tau)$ $(t \geq 0)$ is a semitrajectory for every $\tau \in J$. A set $X_0 \subset X$ is called *invariant* if through each of its points passes at least one trajectory that is entirely contained in X_0. An example of a closed invariant set is the closure of a trajectory.

A set $X_0 \subset X$ is called *minimal* if it is closed, invariant, and does not contain proper closed invariant subsets.

Birkhoff's theorem. *If a semigroup has a compact semitrajectory, then there exists a compact minimal set.*

Proof. Let X_1 denote the closure of a compact semitrajectory. Obviously, the set $\bigcap_{t \geq 0} S(t)X_1$ is compact and invariant. We order the compact invariant sets by inclusion and apply Zorn's lemma, thus proving the existence of a minimal compact invariant set.

The trajectories that belong to a compact minimal set are conventionally called *recurrent* (in the sense of Birkhoff); an example of a recurrent trajectory is an almost periodic trajectory.

2. Suppose that we are given two semigroups defined on X and Y, respectively. Then there is an obvious semigroup on the cartesian product $X \times Y$ (the 'semigroup product').

Two trajectories $x(t)$, $y(t)$ are called *compatibly recurrent* if the trajectory $\{x(t), y(t)\}$ is recurrent in $X \times Y$. Clearly, compatible recurrence implies the recurrence of each component, but the converse does not hold.

We say that a trajectory is *absolutely recurrent* if it is compatibly recurrent with any recurrent trajectory. In Chapter 7 we prove that an almost periodic trajectory is absolutely recurrent.

5 A theorem of A. A. Markov

We consider a semigroup $S(t)$ $(t \geqslant 0)$ on a complete metric space X, and call $S(t)$ *Lyapunov stable* if the transformations $S(t)$ $(t \geqslant 0)$ are equicontinuous on every compact set from X.

Markov's theorem. *The restriction of a Lyapunov stable semigroup to a compact invariant subset is a two-sidedly stable group. In particular, every continuous compact trajectory is almost periodic.*
Proof. Let X be a compact invariant subset. We introduce on X the equivalent metric

$$d(x_1, x_2) = \sup_{t \geqslant 0} \rho(x_1^t, x_2^t),$$

which has the property $d(x_1^t, x_2^t) \leqslant d(x_1, x_2)$ for $t \geqslant 0$. Let $Z = X \times X$. We define a metric on Z by the relation

$$d(z_1, z_2) = d(x_1, x_2) + d(y_1, y_2),$$

where $z_1 = \{x_1, y_1\}$ and $z_2 = \{x_2, y_2\}$. Since X is invariant, through every point $z = z(0) \in Z$ at least one trajectory $z(t)$ passes. Let $A \subset Z$ be the set of elements $z = \{x, y\}$ such that there is at least one trajectory $z(t) = \{x(t), y(t)\}$ with

$$d(x(t), y(t)) \equiv d(x, y) \quad (t \in J).$$

The set A is closed and invariant in Z. We are going to prove that $A = Z$. Suppose this is not the case, that is, there is a $z_0 \notin A$.

Let $z_0(t) = z_0^t$ be some trajectory and z_1 be a limit point of the form

$$z_1 = \lim_{t_m \to -\infty} z_0(t_m).$$

Since the distance $d(z_0, A) > 0$ and the function $d(z_0(t), A)$ is non-increasing, z_1 also does not belong to A. We extract from the sequence $\{t_m\}$ a subsequence $\{t'_m\}$ for which the sequence $z_0(t'_m + \tau)$ is fundamental for any rational τ. Because a translation on Z is continuous for $t \geqslant 0$, the sequence $z_0(t'_m + \tau + \eta)$ is a fundamental sequence for every $\eta \geqslant 0$, that is, $z_0(t'_m + t)$ is a fundamental sequence for every $t \in J$. Therefore we have convergence to some trajectory $z_1(t)$:

$$z_0(t + t'_m) \to z_1(t) \quad (t \in J).$$

Since the function $d(x_0(t), y_0(t))$ is non-increasing, we have

$$d(x_1(t), y_1(t)) = \lim_{m\to\infty} d(x_0(t+t'_m), y_0(t+t'_m)) \equiv \text{const.},$$

that is, $z_1 \in A$. The contradiction proves that $A = Z$.

From our conclusion that $A = Z$ it follows easily that $S(t_0)x_1 \neq S(t_0)x_2$ for $x_1 \neq x_2 (t \geqslant 0)$, that is, through any point from X a unique trajectory passes. It is also easy to conclude that the mapping $S(t_0): X \to X$ is 'onto', that is, the inverse mappings $S^{-1}(t_0)$ are continuous. This proves the theorem.

The next proposition is proved by a similar argument.

Proposition 1. *Suppose that on a compact metric space K there is defined a non-contractive operator $T: K \to K$, that is,*

$$\rho(Tx_1, Tx_2) \geqslant \rho(x_1, x_2).$$

Then $TK = K$.

6 Some simple properties of trajectories

1. The results of this section are purely subsidiary. We consider some general properties of the so-called continuous semigroups.

A semigroup $S(t)$ $(t \geqslant 0)$ is called *continuous* if every semitrajectory of it is a continuous function $J^+ \to X$, where J^+ denotes the semiaxis $[0, \infty)$.

Proposition 2. *Suppose that $S(t)$ is a continuous semigroup on a compact metric space X. Then when t ranges over a finite interval on the open semiaxis* $(0, \infty)$, *the transformations $S(t)$ are equicontinuous.*

Proof. We set $Z = X \times X$, and consider the space B of all continuous scalar functions $\phi(z)$ on Z, and an obvious semigroup of linear operators on B:

$$\phi^t(z) = \phi(S(t)z)$$

(here $z = (x_1, x_2)$). Since the trajectories are continuous, we easily see that the function $\phi^t: J^+ \to B$ is measurable. But then, as is well known from the theory of semigroups of linear operators (see Dunford & Schwartz [40], p. 616), the function ϕ^t is continuous on $(0, \infty)$. Hence, by putting $\phi(z) = \rho(x_1, x_2)$ we obtain the required result.

It follows from Proposition 2 that a compact semitrajectory of a continuous semigroup is uniformly continuous on the semiaxis J^+, and that trajectories belonging to a compact invariant set are uniformly continuous on the whole axis.

2. To the concept of a recurrent trajectory (see § 4) there corresponds the obvious concept of a recurrent function.

Let K be a complete metric space, and let $\Phi(K)$ denote the set of all continuous functions $J \to K$ with the topology of uniform convergence on each finite segment. For $f(s) \in \Phi(K)$ we set $f^t = f(s+t)$.

A function $f(s) \in \Phi(K)$ is called *recurrent* if the trajectory f^t is recurrent in $\Phi(K)$.

There is a natural connection between recurrent functions and recurrent trajectories. Let x^t be the recurrent trajectory of a continuous semigroup defined on a complete metric space X, and let $\phi: X \to K$ be a given continuous function. Then it follows easily from Proposition 2 that $f(t) = \phi(x^t)$ is recurrent. In particular, if ϕ is a scalar function, then since every semitrajectory is everywhere dense in a minimal set, it follows that

$$\sup_{t \in J} f(t) = \sup_{t \leq 0} f(t) = \sup_{t \geq 0} f(t).$$

Comments and references to the literature

§ 1. The definition of an almost periodic function and its simplest properties for numerical functions is due to Bohr [17] and [22]. Long before the publication of Bohr's work, Bohl [15] and Esclangon [120], [121] had discussed a special case of almost periodic functions which are now known as conditionally periodic (or sometimes, quasiperiodic) functions. In contrast to Bohr's definition in which the only condition on almost periods was relative denseness, the definition of Bohl and Esclangon imposed further conditions. The latter definition is as follows: A continuous function f is called conditionally periodic with periods $2\pi/\lambda_1, 2\pi/\lambda_2, \ldots, 2\pi/\lambda_m$ if for every $\varepsilon > 0$ there is a $\delta = \delta(\varepsilon) > 0$ such that each number τ satisfying the system of inequalities

$$|\lambda_k \tau| < \delta \ (\text{mod } 2\pi), \quad k = 1, 2, \ldots, m,$$

also satisfies the inequality

$$\sup_{t \in J} \|f(t+\tau) - f(t)\| \leq \varepsilon,$$

that is, it is an ε-almost period for $f(t)$. The position of conditionally periodic functions in the class of continuous almost periodic functions is discussed in Chapter 3, § 3, and the role of the system of inequalities (8) in the theory of almost periodic functions is considered in Chapter 3, § 2, and Chapter 4, § 1 (Bogolyubov's theorem). The extension of the theory of almost periodic functions to vector-valued (abstract) functions is due to Bochner [27]. Bochner's work was preceded by an important article by Muckenhoupt [93] who considered essentially a special class of abstract almost periodic functions with values in a special Hilbert space. It is interesting to note that the concept of a Bochner measurable and summable function, which is being widely extended at the present time, had its origins in Bochner's investigations on abstract almost periodic functions.

§ 2. The compactness property of an almost periodic function was discovered by Bochner [25]. Lyusternik's theorem was published in [83].

§§ 3 and 4. The connection between almost periodicity and stable dynamical systems is well known (see the monograph of Nemytskii & Stepanov [95], Ch. 5); we have presented only the most elementary facts.

§ 5. The following result is due to Markov: a compact Lyapunov stable trajectory of a dynamical system is almost periodic. In fact, the result which we have called Markov's theorem says slightly more. Proposition 1 was first stated in a paper by Brodskii & Mil'man [30], and then in a more general form in a book by Dunford & Schwartz ([40], p. 459).

2 Harmonic analysis of almost periodic functions

1 Prerequisites about Fourier–Stieltjes integrals

1. Let $\sigma(\lambda)$, $\lambda \in J$, be a numerical (complex-valued) function of bounded variation on the real line.

The *Fourier–Stieltjes transform* of $\sigma(\lambda)$ is the function $f(t)$ defined by

$$f(t) = \int_{-\infty}^{+\infty} \exp(\mathrm{i}\lambda t)\, \mathrm{d}\sigma(\lambda).$$

Let $\lambda_1, \lambda_2, \ldots$ denote the points of discontinuity of $\sigma(\lambda)$ in any order, and $d_1, d_2, \ldots$ the corresponding jumps, that is,

$$d_\nu = \sigma(\lambda_\nu + 0) - \sigma(\lambda_\nu - 0) \quad (\nu = 1, 2, \ldots).$$

We set

$$d(\lambda) = \sum_{\lambda_\nu < \lambda} d_\nu$$

and

$$s(\lambda) = \sigma(\lambda) - d(\lambda).$$

Using this decomposition we can correspondingly represent $f(t)$ as a sum:

$$\begin{aligned} f(t) &= \int_{-\infty}^{+\infty} \exp(\mathrm{i}\lambda t)\, \mathrm{d}d(\lambda) + \int_{-\infty}^{+\infty} \exp(\mathrm{i}\lambda t)\, \mathrm{d}s(\lambda) \\ &= \sum_\nu d_\nu \exp(\mathrm{i}\lambda_\nu t) + \int_{-\infty}^{+\infty} \exp(\mathrm{i}\lambda t)\, \mathrm{d}s(\lambda). \end{aligned} \tag{1}$$

Lemma 1

$$\lim_{T\to\infty} \frac{1}{2T} \int_{-T}^{T} f(t) \exp(-\mathrm{i}\mu t)\, \mathrm{d}t = \begin{cases} d_\nu & \text{if } \mu = \lambda_\nu, \\ 0 & \text{if } \mu \neq \lambda_\nu \end{cases} \quad (\nu = 1, 2, \ldots). \tag{2}$$

Proof. It follows from (1) that

$$\frac{1}{2T}\int_{-T}^{T} f(t)\exp(-\mathrm{i}\mu t)\,\mathrm{d}t$$

$$=\begin{cases} d_{\nu_0}+\sum\limits_{\nu\neq\nu_0} d_\nu \dfrac{\sin(\lambda_\nu-\lambda_{\nu_0})T}{T(\lambda_\nu-\lambda_{\nu_0})} \\ \quad +\displaystyle\int_{-\infty}^{\infty}\frac{\sin(\lambda-\lambda_{\nu_0})T}{T(\lambda-\lambda_{\nu_0})}\,\mathrm{d}s(\lambda) \quad \text{if } \mu=\lambda_{\nu_0}, \\ \sum\limits_{\nu} d_\nu \dfrac{\sin(\lambda_\nu-\mu)T}{T(\lambda_\nu-\mu)} \\ \quad +\displaystyle\int_{-\infty}^{\infty}\frac{\sin(\lambda-\mu)T}{T(\lambda-\mu)}\,\mathrm{d}s(\lambda) \quad \text{if } \mu\neq\lambda_{\nu_0} \quad (\nu=1,2,\ldots). \end{cases} \tag{3}$$

From (3) it follows that (2) will be proved if we can establish that

$$\lim_{T\to\infty}\int_{-\infty}^{\infty}\frac{\sin(\lambda-\mu)T}{T(\lambda-\mu)}\,\mathrm{d}s(\lambda)=0 \tag{4}$$

for all $\mu\in J$, and that for $\mu\neq\lambda_\nu$, $\nu=1,2,\ldots,$

$$\lim_{T\to\infty}\sum_{\nu} d_\nu\frac{\sin(\lambda_\nu-\mu)T}{T(\lambda_\nu-\mu)}=0. \tag{5}$$

The series in (5) is majorised by $\sum_\nu |d_\nu|/T|\lambda_\nu-\mu|$, and for $\mu\neq\lambda_\nu$ $(\nu=1,2,\ldots)$ each term in the series (5) tends to zero as $T\to\infty$; hence we obtain the equality (5).

For the proof of (4) we set for any $\delta>0$

$$\begin{aligned}\int_{-\infty}^{\infty}\frac{\sin(\lambda-\mu)T}{(\lambda-\mu)T}\,\mathrm{d}s(\lambda) &= \int_{-\infty}^{\infty}\frac{\sin\lambda T}{\lambda T}\,\mathrm{d}_\lambda s(\lambda+\mu) \\ &= \int_{|\lambda|\leq\delta}\frac{\sin\lambda T}{\lambda T}\,\mathrm{d}_\lambda s(\lambda+\mu) \\ &\quad +\int_{|\lambda|>\delta}\frac{\sin\lambda T}{\lambda T}\,\mathrm{d}_\lambda s(\lambda+\mu) \\ &= A_1+A_2.\end{aligned}$$

Then by using standard estimates for Riemann–Stieltjes integrals we have

$$\begin{aligned}|A_1| &\leq \int_{-\delta}^{\delta}|\mathrm{d}_\lambda s(\lambda+\mu)| = \mathrm{Var}_{\mu-\delta}^{\mu+\delta}\{s(\lambda)\} \\ &= \mathrm{Var}_0^{\mu+\delta}\{s(\lambda)\}-\mathrm{Var}_0^{\mu-\delta}\{s(\lambda)\}; \\ |A_2| &\leq \frac{1}{\delta T}\mathrm{Var}_{-\infty}^{\infty}\{s(\lambda)\}.\end{aligned}$$

The relation (4) now follows from these estimates and from the continuity of the variation of a continuous function of bounded variation.[1]

2. Now suppose that $g(t)$ is the Fourier–Stieltjes transform of a function $\tau(\lambda)$ (also of bounded variation on the whole line):

$$g(t)=\int_{-\infty}^{\infty} \exp(\mathrm{i}\lambda t)\,\mathrm{d}\tau(\lambda).$$

We consider the product $f(t)\cdot g(t)$; it is a standard result that it can also be represented as a Fourier–Stieltjes integral:[2]

$$f(t)\cdot g(t)=\int_{-\infty}^{\infty} \exp(\mathrm{i}\lambda t)\,\mathrm{d}\rho(\lambda),$$

where

$$\rho(\lambda)=\int_{-\infty}^{\infty} \sigma(\lambda-\mu)\,\mathrm{d}\tau(\mu)=\int_{-\infty}^{\infty} \sigma(\mu)\,\mathrm{d}_\mu\tau(\lambda-\mu). \tag{6}$$

The integral in (6) is called the *convolution* of the functions σ and τ. From (6) it follows that if at least one of $\sigma(\lambda)$ or $\tau(\lambda)$ is continuous, then so is their convolution.

Now we consider a special case. Let

$$h(t)=\int_{-\infty}^{\infty} \exp(\mathrm{i}\lambda t)\,\mathrm{d}s(\lambda), \tag{7}$$

where $s(\lambda)$ is continuous. It is easy to see that

$$\overline{h(t)}=\int_{-\infty}^{\infty} \exp(-\mathrm{i}\lambda t)\,\mathrm{d}\overline{s(\lambda)}=\int_{-\infty}^{\infty} \exp(\mathrm{i}\lambda t)\,\mathrm{d}\overline{s(-\lambda)}.$$

Hence, it follows from (6) that the function

$$|h(t)|^2=h(t)\cdot\overline{h(t)}$$

is representable as a Fourier–Stieltjes integral with a continuous distribution function. This conclusion and Lemma 1 (with $\mu=0$) lead to

Lemma 2. *Let $h(t)$ be represented as a Fourier–Stieltjes integral* (7) *with a continuous distribution function $s(\lambda)$. Then*

$$\lim_{T\to\infty}\frac{1}{2T}\int_{-T}^{T} |h(t)|^2\,\mathrm{d}t=0.$$

1 G. E. Shilov, *Mathematical analysis. A special course*, Moscow, 1961, p. 280.

2 A. N. Kolmogorov & S. V. Fomin, *Elements of the theory of functions and functional analysis*, 'Nauka', Moscow, 1972, p. 423.

3. Now we assume that the distribution function $\sigma(\lambda)$ is non-decreasing and bounded for $-\infty<\lambda<\infty$. This case often occurs in various applications. Let

$$f(t)=\int_{-\infty}^{\infty} \exp(i\lambda t)\, d\sigma(\lambda), \tag{8}$$

$t_1, t_2, \ldots, t_n$ be arbitrary real numbers, and $\xi_1, \xi_2, \ldots, \xi_n$ be arbitrary complex numbers. From (8) it follows that

$$\begin{aligned}
&\sum_{\mu=1}^{n}\sum_{\nu=1}^{n} f(t_\mu - t_\nu)\xi_\mu\bar{\xi}_\nu \\
&\quad=\int_{-\infty}^{\infty}\left[\sum_{\mu=1}^{n}\sum_{\nu=1}^{n}\xi_\mu\bar{\xi}_\nu \exp(i\lambda t_\mu)\exp(-i\lambda t_\nu)\right] d\sigma(\lambda)\\
&\quad=\int_{-\infty}^{\infty}\left|\sum_{\mu=1}^{n}\xi_\mu \exp(i\lambda t_\mu)\right|^2 d\sigma(\lambda)\geqslant 0.
\end{aligned}$$

Definition. A function $f: J\to C$ for which

$$\sum_{\mu=1}^{n}\sum_{\nu=1}^{n} f(t_\mu - t_\nu)\xi_\mu\bar{\xi}_\nu \geqslant 0$$

for arbitrary real numbers $t_1, t_2, \ldots, t_n$ and arbitrary complex numbers $\xi_1, \xi_2, \ldots, \xi_n$, is called *positive-definite.*

We need the following classical theorem of Bochner and Khinchin, which we state without proof.[3]

Every continuous positive-definite function f can be represented (in a unique way) as an integral (8) *with a non-decreasing bounded function $\sigma(\lambda)$.*

Remark. In an expansion (1) for a positive-definite function we have $d_\nu>0$ and $s(\lambda)$ is a continuous non-decreasing function.

2 Proof of the approximation theorem

1. The approximation theorem. *For every continuous almost periodic function $f: J\to X$ and for every $\varepsilon>0$ there is a trigonometric polynomial*

$$P_\varepsilon(t)=\sum_{\nu=1}^{N_\varepsilon} b_{\nu,\varepsilon}\exp(i\lambda_{\nu,\varepsilon}t) \quad (b_{\nu,\varepsilon}\in X, \lambda_{\nu,\varepsilon}\in J)$$

3 It is proved, for example, in the book by G. E. Shilov (referred to in footnote 1), p. 404.

such that

$$\sup_{t\in J} \|f(t)-P_\varepsilon(t)\|<\varepsilon.$$

Remark 1. We can deduce from the approximation theorem other basic theorems in the theory of almost periodic functions, and so it must be regarded as a principal result of this theory.

Remark 2. The function $a\exp(\mathrm{i}\lambda t)$ is periodic of period $2\pi/|\lambda|$ for all $a\in X$ and for all $\lambda\in J$. Property 6 (Chapter 1, § 2) implies that every trigonometric polynomial is an almost periodic function, and it follows from Property 3 (Chapter 1, § 4) that every uniform limit of trigonometric polynomials is an almost periodic function. The approximation theorem asserts that all almost periodic functions can be obtained in this way.

Proof of the approximation theorem. For any $\varepsilon>0$ we choose numbers $l=l(\varepsilon/8)$ and $\delta=\delta(\varepsilon/8)$ in accordance with Definition 3 (Chapter 1, § 1) and Property 2 (Chapter 1, § 1), respectively. Then in any interval of length l there is a number τ such that

$$\sup_{t\in J}\|f(t+\tau)-f(t)\|\leqslant\varepsilon/8 \tag{9}$$

and

$$\|f(t'')-f(t')\|\leqslant\varepsilon/8 \tag{10}$$

for any $t', t''\in J$ with $|t''-t'|<\delta$. We cover J with the intervals

$$J_n=(nl,(n+1)l),\quad n=0,\pm1,\pm2,\ldots$$

From (9) and (10) it follows that each interval J_n contains a subinterval $\Delta_n=(\tau_n-\delta,\tau_n+\delta)$ whose points are $(\varepsilon/4)$-almost periods of $f(t)$, that is, for every $\tau'\in\Delta_n$

$$\sup_{t\in J}|f(t+\tau')-f(t)|\leqslant\varepsilon/4. \tag{11}$$

We define a function $K_\delta(s)$, $s\in J$, by

$$K_\delta(s)=\begin{cases} l/2\delta & \text{for } s\in\Delta_n,\\ 0 & \text{for } s\notin\Delta_n.\end{cases}$$

It has the following obvious properties, which we shall need later on:

(1) $(1/2nl)\int_{-nl}^{nl}K_\delta(s)\,\mathrm{d}s=1 \quad (n=1,2,\ldots)$.

(2) For any $s\in J$ and any natural number m

$$\frac{1}{2ml}\int_{-ml+s}^{ml+s}K_\delta(r)\,\mathrm{d}r=1+\eta(s),$$

where

$$|\eta(s)| \leq 1/m.$$

(3) For every fixed δ the set of functions

$$\phi_{\delta,T}(u) = \frac{1}{2T}\int_{-T}^{T} K_\delta(s)K_\delta(u+s)\,\mathrm{d}s,$$

$$T = T_n = nl, \quad n = 1, 2, \ldots, \quad -\infty < u < \infty,$$

is uniformly bounded and equicontinuous. Hence, by the classical theorem of Arzela we can find a subsequence $T_k = T_{n_k}$ of the sequence T_n such that the limit

$$\phi_\delta(u) = \lim_{k\to\infty}\frac{1}{2T_k}\int_{-T_k}^{T_k} K_\delta(s)K_\delta(u+s)\,\mathrm{d}s \quad (-\infty < u < \infty)$$

exists uniformly in every finite interval.

(4) The limit function $\phi_\delta(u)$ is positive definite. For

$$\sum_{\mu=1}^{n}\sum_{\nu=1}^{n}\phi_\delta(u_\mu - u_\nu)\xi_\mu\bar{\xi}_\nu$$

$$= \sum_{\mu=1}^{n}\sum_{\nu=1}^{n}\lim_{k\to\infty}\left\{\frac{1}{2T_k}\int_{-T_k}^{T_k} K_\delta(s)K_\delta(u_\mu - u_\nu + s)\xi_\mu\bar{\xi}_\nu\,\mathrm{d}s\right\}$$

$$= \sum_{\mu=1}^{n}\sum_{\nu=1}^{n}\left\{\lim_{k\to\infty}\frac{1}{2T_k}\int_{-T_k}^{T_k} K_\delta(u_\nu + s)K_\delta(u_\mu + s)\xi_\mu\bar{\xi}_\nu\,\mathrm{d}s\right\}$$

$$= \lim_{k\to\infty}\frac{1}{2T_k}\int_{-T_k}^{T_k}\left|\sum_{\mu=1}^{n} K_\delta(u_\mu + s)\xi_\mu\right|^2 \mathrm{d}s \geq 0.$$

From the Bochner–Khinchin theorem we obtain the representation

$$\phi_\delta(u) = \sum_{\nu=1}^{\infty}\alpha_\nu \exp(\mathrm{i}\lambda_\nu t) + \int_{-\infty}^{\infty}\exp(\mathrm{i}\lambda t)\,\mathrm{d}s(\lambda), \tag{12}$$

where $\alpha_\nu > 0$, and $s(\lambda)$ is continuous, non-decreasing and bounded.

2. Next we choose natural numbers m and n arbitrarily and set

$$f_{\delta,m,n}(t) = \frac{1}{4mnl^2}\int_{-nl}^{nl}\left\{\int_{-ml+s}^{ml+s} K_\delta(s)K_\delta(r)f(t-s+r)\,\mathrm{d}r\right\}\mathrm{d}s.$$

From (11) we see that if $K_\delta(s)K_\delta(r) \neq 0$, then $-s+r$ is an $(\varepsilon/2)$-almost period of f. Therefore, it follows from properties (1) and (2) of $K_\delta(s)$ that

$$\sup_{t\in J}\|f(t) - f_{\delta,m,n}(t)\| \leq \frac{\varepsilon}{2} + \frac{\Gamma}{m}, \tag{13}$$

where

$$\Gamma = \sup_{t\in J} \|f(t)\|.$$

We set $nl = T$ and $ml = R$ to give

$$f_{\delta,m,n}(t)$$
$$= \frac{1}{2T}\int_{-T}^{T}\left\{\frac{1}{2R}\int_{-R+s}^{R+s} K_\delta(s)K_\delta(r)f(t-s+r)\,\mathrm{d}r\right\}\mathrm{d}s$$
$$= \frac{1}{2T}\int_{-T}^{T}\left\{\frac{1}{2R}\int_{-R}^{R} K_\delta(s)K_\delta(u+s)f(t+u)\,\mathrm{d}u\right\}\mathrm{d}s.$$

Thus by setting $T = T_k$ (see property (4) in the preceding subsection) and taking the limit, after using properties (3) and (4) of $K_\delta(s)$ and the equality (12), we obtain

$$f_{\delta,m}(t) \stackrel{\text{def}}{=} \lim_{k\to\infty} f_{\delta,m,m_k}(t)$$
$$= \frac{1}{2R}\int_{-R}^{R} f(t+u)\phi_\delta(u)\,\mathrm{d}u$$
$$= \sum_\nu \alpha_\nu \exp(-\mathrm{i}\lambda_\nu t)\frac{1}{2R}\int_{-R}^{R} f(t+u)$$
$$\times \exp[\mathrm{i}\lambda(t+u)]\,\mathrm{d}u$$
$$+\frac{1}{2R}\int_{-R}^{R} f(t+u)h(u)\,\mathrm{d}u,$$

where

$$h(u) = \int_{-\infty}^{\infty} \exp(\mathrm{i}\lambda u)\,\mathrm{d}s(\lambda).$$

Hence, from (13) with $n = m_k$, in the limit as $k\to\infty$ we obtain

$$\|f(t) - f_{\delta,m}(t)\| \leq \varepsilon/2 + \Gamma/m. \tag{14}$$

To complete the proof of the approximation theorem we need the following simple lemma.

Lemma 3. *Suppose that a function $f: J\to X$ has a compact trajectory. Then the function*

$$F(T) = \frac{1}{2T}\int_{-T}^{T} f(t)\,\mathrm{d}t \quad (T > 1)$$

also has a compact trajectory.

Proof. It is easy to see from the definition of a Riemann integral that $F(T)$ belongs to the closed convex hull of $\mathscr{R}_f$. But, as is well known, the latter is compact together with $\mathscr{R}_f$.

Now we can complete the proof of the approximation theorem. From Lemma 2 it follows that

$$\begin{aligned}\overline{\lim_{R\to\infty}} &\left\|\frac{1}{2R}\int_{-R}^{R} f(t+u)h(u)\,\mathrm{d}u\right\| \\ &\leqslant \overline{\lim_{R\to\infty}}\left(\frac{1}{2R}\int_{-R}^{R}\|f(t+u)\|^2\,\mathrm{d}u\right)^{1/2} \\ &\quad\times \lim_{R\to\infty}\frac{1}{2R}\int_{-R}^{R}|h(u)|^2\,\mathrm{d}u = 0. \end{aligned} \tag{15}$$

Thus, if $\alpha_1, \alpha_2, \ldots, \alpha_N$ are chosen so that

$$\sum_{\nu=N+1}^{\infty} \alpha_\nu < \varepsilon/2\Gamma,$$

then for all $R>0$ we shall have

$$\left\|\sum_{\nu=N+1}^{\infty} \alpha_\nu \frac{1}{2R}\int_{-R}^{R} f(t+u)\exp\left[\mathrm{i}\lambda(t+u)\right]\mathrm{d}u\right\| < \varepsilon/2. \tag{16}$$

By Lemma 3 there is a subsequence $R_k = m_k l$ such that the following limits exist for all $\nu = 1, 2, \ldots, N$:

$$\begin{aligned} A_\nu &\overset{\text{def}}{=} \lim_{k\to\infty}\frac{1}{2R_k}\int_{-R_k}^{R_k} f(t+u)\exp\left[\mathrm{i}\lambda_\nu(t+u)\right]\mathrm{d}u \\ &= \lim_{k\to\infty}\frac{1}{2R_k}\int_{-R_k}^{R_k} f(t)\exp\left(\mathrm{i}\lambda_\nu t\right)\mathrm{d}t. \end{aligned} \tag{17}$$

It follows from (14), (15), and (16) that

$$\|f(t) - \sum A_\nu \alpha_\nu \exp(-\mathrm{i}\lambda_\nu t)\| < \varepsilon,$$

and this completes the proof of the approximation theorem.

3 The mean-value theorem; the Bohr transformation; Fourier series; the uniqueness theorem

1. We make the important point that in proving the approximation theorem we have used only the definition of an almost periodic function and the elementary Property 2 in Chapter 1. On the other hand, as we are going to show in this section, basic properties of an almost periodic function can be deduced comparatively simply from the approximation theorem. Clearly, there is no need to derive once again the properties we have already mentioned, but as an example we consider the theorem about a sum of almost periodic functions. Let $f(t)$, $g(t)$ be two almost periodic functions. For every $\varepsilon>0$ there are trigonometric polynomials $P_{\varepsilon/2}(t)$ and $Q_{\varepsilon/2}(t)$ such that

$$\sup_{t\in J}\|f(t)-P_{\varepsilon/2}(t)\|<\varepsilon/2,$$

$$\sup_{t\in J}\|g(t)-Q_{\varepsilon/2}(t)\|<\varepsilon/2.$$

Consequently,

$$\sup_{t\in J}\|f(t)+g(t)-[P_{\varepsilon/2}(t)+Q_{\varepsilon/2}(t)]\|<\varepsilon,$$

and so $f(t)+g(t)$ is an almost periodic function (see Chapter 1, § 1, Property 3). We show that other basic theorems in the theory of almost periodic functions can also be deduced from the approximation theorem.

2. Property 1. The mean-value theorem. *For every almost periodic function $f(t)$, the mean value*

$$\lim_{T\to\infty}\frac{1}{2T}\int_{-T}^{T}f(t)\,\mathrm{d}t\overset{\text{def}}{=}\boldsymbol{M}\{f\}$$

exists. In addition, the limit

$$\lim_{T\to\infty}\frac{1}{2T}\int_{-T+a}^{T+a}f(t)\,\mathrm{d}t=\boldsymbol{M}\{f\} \tag{18}$$

exists uniformly with respect to a.

Proof. For all $a\in J$ and for all $T>0$

$$\frac{1}{2T}\int_{-T+a}^{T+a}\exp(\mathrm{i}\lambda t)\,\mathrm{d}t=\begin{cases}1 & \text{if } \lambda=0,\\ \exp(\mathrm{i}\lambda a)\dfrac{\sin\lambda T}{T} & \text{if } \lambda\neq 0.\end{cases}$$

Therefore, for all $\lambda\in J$,

$$\lim_{T\to\infty}\frac{1}{2T}\int_{-T+a}^{T+a}\exp(\mathrm{i}\lambda t)\,\mathrm{d}t\overset{\text{def}}{=}\psi(\lambda)=\begin{cases}1 & \text{if } \lambda=0,\\ 0 & \text{if } \lambda\neq 0\end{cases} \tag{19}$$

uniformly with respect to a. From (19) it follows that (18) holds uniformly in a for every trigonometric polynomial

$$P(t)=\sum_{k=1}^{N}a_k\exp(\mathrm{i}\lambda_k t)\quad(a_k\in X),$$

and

$$\boldsymbol{M}\{P(t)\}=\sum_{k=1}^{N}a_k\psi(\lambda_k).$$

Next let f be an arbitrary almost periodic function; then for every $\varepsilon > 0$ there is a trigonometric polynomial $P_\varepsilon(t)$ such that

$$\sup_{t \in J} \|f(t) - P_\varepsilon(t)\| < \varepsilon.$$

We set

$$\frac{1}{2T} \int_{-T+a}^{T+a} f(t)\, \mathrm{d}t \overset{\text{def}}{=} \boldsymbol{M}\{f; T, a\}.$$

Then we have

$$\begin{aligned} &\|\boldsymbol{M}\{f; T', a\} - \boldsymbol{M}\{f; T'', a\}\| \\ &\quad \leqslant \|\boldsymbol{M}\{f - P_\varepsilon; T', a\}\| + \|\boldsymbol{M}\{P_\varepsilon; T', a\} \\ &\qquad - \boldsymbol{M}\{P_\varepsilon; T'', a\}\| + \|\boldsymbol{M}\{P_\varepsilon - f; T'', a\}\| \\ &\quad \leqslant \|\boldsymbol{M}\{P_\varepsilon; T', a\} - \boldsymbol{M}\{P_\varepsilon; T'', a\}\| + 2\varepsilon. \end{aligned}$$

Therefore, for every $\varepsilon > 0$ there exists a $T_\varepsilon > 0$ such that for $T', T'' > T_\varepsilon$ and for any $a \in J$ we have

$$\|\boldsymbol{M}\{f; T', a\} - \boldsymbol{M}\{f; T'', a\}\| < 3\varepsilon,$$

as we required to prove. Notice that for any almost periodic function f and for all $\lambda \in J$ the function $f(t) \exp(-\mathrm{i}\lambda t)$ is almost periodic. Hence, the function

$$a(\lambda; f) = \boldsymbol{M}_t\{f(t) \exp(-\mathrm{i}\lambda t)\}$$

is defined for every $\lambda \in J$; $a(\lambda; f)$ is called the *Bohr transformation* of f.

The next property is fundamental to the theory of almost periodic functions.

Property 2. *The function $a(\lambda; f)$ is non-zero for at most a countable set of values of λ.*

To prove this property let $\{P_k(t)\}$ be a sequence of approximating polynomials for which

$$\sup_{t \in J} \|f(t) - P_k(t)\| \leqslant 1/k \quad (k = 1, 2, \ldots).$$

Suppose that

$$P_k(t) = \sum_{m=1}^{n_k} a_{k,m} \exp(\mathrm{i}\lambda_{k,m} t)$$

and

$$\{\mu_n\} = \bigcup_{k,m} \lambda_{k,m}.$$

The set $\{\mu_n\}$ is not more than countable. We show that $a(\lambda; f) = 0$ for $\lambda \neq \mu_n$. In fact,

$$\|a(\lambda; f)\| = \|a(\lambda; P_k) + a(\lambda; f - P_k)\|$$
$$= \|a(\lambda; f - P_k)\| \leqslant 1/k \quad (k = 1, 2, \ldots).$$

Since we can take k arbitrarily large, we conclude that $a(\lambda; f) = 0$ for $\lambda \neq \mu_n$.

The set $\{\lambda_n\}$ of all those λ for which $a(\lambda; f) \neq 0$ is called the *spectrum* of f; obviously, $\{\lambda_n\} \subseteq \{\mu_n\}$. Let

$$a_n = a(\lambda_n; f).$$

With each almost periodic function f we associate (formally, for the time being) the Fourier series

$$f(t) \sim \sum_n a_n \exp(i\lambda_n t).$$

($\sim$ means that there is a relation between the a_n and $f(t)$ and conveys no implication of convergence.) The elements $a_n \in X$ are called the *Fourier coefficients* and the numbers $\{\lambda_n\}$ the *Fourier exponents* of f. The next property follows easily from the proof of the approximation theorem and from the mean-value theorem (see (18)).

Property 3. *The Fourier exponents of approximating polynomials*

$$P_\varepsilon(t) = \sum_{k=1}^{n_\varepsilon} a_{k,\varepsilon} \exp(i\lambda_{k,\varepsilon} t)$$

can be chosen from those of $f(t)$; the Fourier coefficients of approximating polynomials can be regarded as the products of the Fourier coefficients of the function and certain positive numbers (depending on ε and the Fourier exponents of the function).

As a simple consequence of Property 3 we obtain the following important result.

The uniqueness theorem. *Let $f(t)$ and $g(t)$ be two almost periodic functions. If $a(\lambda; f) \equiv a(\lambda; g)$, then $f \equiv g$.*

Proof. If $a(\lambda; f) \equiv a(\lambda; g)$, then $a(\lambda; f - g) \equiv 0$. Therefore, we can assume that the approximating polynomial $P_\varepsilon(t; f - g) \equiv 0$ for every $\varepsilon > 0$. Consequently, $f(t) - g(t) \equiv 0$.

Property 4. *For any almost periodic function f we have* $\lim_{n\to\infty} a_n = 0$.

In fact, let $P_\varepsilon(t)$ be a trigonometric polynomial for which

$$\sup_{t \in J} \|f(t) - P_\varepsilon(t)\| \leqslant \varepsilon,$$

and n_ε be such that $\boldsymbol{M}\{P_\varepsilon(t)\exp(-i\lambda_n t)\}=0$ for $n>n_\varepsilon$. Then for $n>n_\varepsilon$

$$\begin{aligned}\|a_n\| &= \|\boldsymbol{M}\{f(t)\exp(-i\lambda_n t)\}\| \\ &= \|\boldsymbol{M}_t\{[f(t)-P_\varepsilon(t)]\exp(-i\lambda_n t)\}\| \leqslant \varepsilon.\end{aligned}$$

4 Bochner–Fejer polynomials

1. Let f be a 2π-periodic function with a Fourier series

$$f(t) \sim \sum_{k=-\infty}^{+\infty} a_k \exp(ikt),$$

$$a_k = \frac{1}{2\pi}\int_{-\pi}^{\pi} f(t)\exp(-ikt)\,dt.$$

We set

$$s_n(t) = \sum_{k=-n}^{n} a_k \exp(ikt), \quad s_0(t) = a_0,$$

and call $s_n(t)$ a partial sum (segment) of the Fourier series. Next we write

$$\begin{aligned}\sigma_n(t) &= \frac{s_0(t)+s_1(t)+\cdots+s_{n-1}(t)}{n} \\ &= \sum_{k=-n}^{n}\left[1-\frac{|k|}{n}\right] a_k \exp(ikt).\end{aligned}$$

The sums $\sigma_n(t)$ are called *Fejer sums*. For every continuous periodic function f, the Fejer sums converge uniformly to $f(t)$ as $n\to\infty$. On the one hand, the Fejer sums are the arithmetic means of the partial sums of the Fourier series of f, and on the other hand they can be obtained from the Fourier series by introducing into the series the multipliers

$$r_k^{(n)} = \begin{cases}\left(1-\dfrac{|k|}{n}\right) & \text{for } |k|<n, \\ 0 & \text{for } |k|\geqslant n.\end{cases}$$

Bochner has proved that one can introduce in a Fourier series of any almost periodic function multipliers depending essentially on the Fourier exponents of the function, so as to obtain finite trigonometric sums that converge uniformly to the almost periodic function. These sums are obvious generalisations of Fejer sums, and so we call them *Bochner–Fejer sums*. The present section is devoted to the construction of Bochner–Fejer sums, and to the proof of their convergence.

2. We assemble some simple concepts that will play a significant role in other questions.

Definition 1. A finite or countable set of real numbers $\beta_1, \beta_2, \ldots, \beta_n, \ldots$ is said to be *linearly independent* (over the field of rational numbers) if the equality

$$r_1\beta_1 + r_2\beta_2 + \cdots + r_n\beta_n = 0$$

($r_1, r_2, \ldots, r_n$ are rational and n is an arbitrary natural number) implies that all of $r_1, r_2, \ldots, r_n$ are zero.

Definition 2. A finite or countable set of linearly independent real numbers $\beta_1, \beta_2, \ldots, \beta_n, \ldots$ is called a *rational basis* of a countable set of real numbers $\lambda_1, \lambda_2, \ldots, \lambda_n, \ldots$ if every λ_n is representable as a finite linear combination of the β_j with rational coefficients, that is,

$$\lambda_n = r_1^{(n)}\beta_1 + r_2^{(n)}\beta_2 + \cdots + r_{m_k}^{(n)}\beta_{m_k} \quad (n = 1, 2, \ldots), \tag{20}$$

where the $r_j^{(n)}$ are rational numbers.

Theorem. *Every countable set of real numbers has a basis contained in the set.*
Proof. Let

$$\lambda_1, \lambda_2, \ldots, \lambda_n, \ldots \tag{21}$$

be a given set of real numbers. We denote the first non-zero number in the set by β_1 and delete from the set (21) all numbers λ satisfying the equality

$$r_1\beta_1 + r_2\lambda = 0$$

for rational numbers r_1 and r_2. Let β_2 be the first number in (21) not deleted in this way. We then remove from the set (21) all those λ for which

$$r_1\beta_1 + r_2\beta_2 + r_3\lambda = 0$$

for rational numbers r_1, r_2, and r_3. By continuing this process we construct a basis for the set (21).
Remark. Clearly, a given set of numbers can have several rational bases, but in a specific basis the representation (20) is unique.

If a basis consists of a finite number of terms, then it is called a *finite basis*, otherwise, it is *infinite*. If in the representation (20) all the $r_j^{(n)}$ are integers, then the basis is called an *integer* basis. A basis can be both finite and integer. For instance, let β_1, β_2 be non-coprime real numbers and consider the countable set of numbers of the form $n_1\beta_1+n_2\beta_2$, where n_1 and n_2 are integers. Obviously, the numbers (β_1, β_2) form an integer basis for this set.

3. Let $f(t)$ be periodic of period $p=2\pi/|\beta|$, and

$$f(t)\sim\sum_{-\infty}^{+\infty} a_k \exp(ik\beta t),$$

where $a_k=(1/p)\int_0^p f(t)\exp(-ik\beta t)\,dt$. We prove that

$$a_k=\boldsymbol{M}_t\{f(t)\exp(-ik\beta t)\}. \tag{22}$$

In fact, for an arbitrary $T>0$ we have

$$\begin{aligned}\frac{1}{2T}&\int_{-T}^{T} f(t)\exp(-ik\beta t)\,dt\\ &=\frac{1}{2T}\left\{\int_{-T}^{-Np} f(t)\exp(-ik\beta t)\,dt\right.\\ &\quad+\int_{-Np}^{Np} f(t)\exp(-ik\beta t)\,dt\\ &\quad\left.+\int_{Np}^{T} f(t)\exp(-ik\beta t)\,dt\right\},\end{aligned} \tag{23}$$

where $N=[T/p]$. Since $f(t)\exp(-ik\beta t)$ is periodic,

$$\int_{-Np}^{Np} f(t)\exp(-ik\beta t)\,dt=2N\int_0^p f(t)\exp(-ik\beta t)\,dt,$$

and so we obtain (22) from (23) by letting $T\to\infty$ $(N\to\infty)$. Let us calculate the Fejer sum of order n of the periodic function f. First we have

$$\begin{aligned}s_r(t)&=\sum_{\nu=-r}^{r} a_\nu \exp(i\nu\beta t)\\ &=\boldsymbol{M}_s\left\{f(s)\sum_{\nu=-r}^{r}\exp[-i\nu\beta(s-t)]\right\}\\ &=\boldsymbol{M}_s\left\{f(t+s)\frac{\sin(r+\frac{1}{2})\beta s}{\sin(\beta s/2)}\right\}\\ &=\boldsymbol{M}_s\left\{f(t+s)\frac{\sin(\beta s/2)\sin(r+\frac{1}{2})\beta s}{\sin^2(\beta s/2)}\right\}\\ &=\boldsymbol{M}_s\left\{f(t+s)\frac{\cos r\beta s-\cos(r+1)\beta s}{\sin^2(\beta s/2)}\right\}.\end{aligned}$$

Therefore,

$$\sigma_n(t) = \frac{s_0(t)+s_1(t)+\cdots+s_{n-1}(t)}{n}$$
$$= \boldsymbol{M}_s\left\{f(t+s)\frac{\sin^2(n\beta s/2)}{n\sin^2(\beta s/2)}\right\}.$$

On the other hand,

$$\sigma_n(t) = \sum_{\nu=-n}^{n}\left(1-\frac{|\nu|}{n}\right) a_\nu \exp(\mathrm{i}\nu\beta t).$$

The function

$$K_n(\beta s) = \frac{\sin^2(n\beta s/2)}{n\sin^2(\beta s/2)}$$
$$= \sum_{\nu=-n}^{n}\left(1-\frac{|\nu|}{n}\right)\exp(-\mathrm{i}\nu\beta s) \tag{24}$$

is called the *Fejer kernel* of *order* n. It has two important properties which follow directly from (24):

(1) $K_n(\beta s) \geq 0$,

(2) $\boldsymbol{M}_s\{K_n(\beta s)\} = 1$.

4. Now we assume that $f: J \to X$ is an almost periodic function with a Fourier series

$$f(t) \sim \sum_n a_n \exp(\mathrm{i}\lambda_n t).$$

Let $\beta_1, \beta_2, \ldots, \beta_r, \ldots$ be a rational basis for the Fourier exponents of $f(t)$, let $\tilde{B}$ be the set of all finite linear combinations of the numbers $\beta_1, \beta_2, \ldots, \beta_r, \ldots$ with rational coefficients, and let m be an arbitrary natural number.

By a *composite* Bochner–Fejer kernel we mean the function

$$K_{m;\beta_1,\beta_2,\ldots,\beta_m}(s) = K_{(m!)^2}\left(\frac{\beta_1 s}{m!}\right)\cdots K_{(m!)^2}\left(\frac{\beta_m s}{m!}\right)$$
$$= \sum_{\substack{|\nu_1|\leq(m!)^2\\ \vdots \\ |\nu_m|\leq(m!)^2}}\left(1-\frac{|\nu_1|}{(m!)^2}\right)\cdots\left(1-\frac{|\nu_m|}{(m!)^2}\right)$$
$$\times\exp\left[\mathrm{i}\left(\frac{\nu_1}{m!}\beta_1+\cdots+\frac{\nu_m}{m!}\beta_m\right)t\right]$$
$$= \sum_{\substack{|\nu_1|\leq(m!)^2\\ \vdots \\ |\nu_m|\leq(m!)^2}} K_{m;\nu_1,\ldots,\nu_m}$$
$$\times\exp\left[\mathrm{i}\left(\frac{\nu_1}{m!}\beta_1+\cdots+\frac{\nu_m}{m!}\beta_m\right)t\right].^4 \tag{25}$$

4 We shall see later that this formula is acceptable also in the case of a finite basis.

Since the numbers $\beta_1, \beta_2, \ldots$ are linearly independent, the term under the exponential in (25) is zero only when $\nu_1 = \nu_2 = \cdots = \nu_m = 0$. Therefore,

$$\boldsymbol{M}_s\{K_{m;\beta_1,\ldots,\beta_m}(s)\} = 1,$$

that is, the Fejer composite kernel has the second property of a simple Fejer kernel. The first property is obvious from the representation of a composite kernel as a product of simple Fejer kernels.

The coefficients $K_{m;\nu_1,\ldots,\nu_m}$ in the last sum (25) have the obvious properties:

(1) They are all positive and are not greater than 1.

(2) $K_{m;0,0,\ldots,0} = 1$.

(3) For fixed $r, \nu_1, \nu_2, \ldots, \nu_r$, as $m \to \infty$,

$$K_{m;\nu_1,\ldots,\nu_r;0,\ldots 0} \to 1.$$

Then for a given almost periodic function $f(t)$ we define a trigonometric polynomial of Bochner–Fejer by

$$\begin{aligned} P_m(t; f) &= \boldsymbol{M}_s\{f(t+s)K_{m;\beta_1,\ldots,\beta_m}(s)\} \\ &= \sum_{\substack{|\nu_1|\leq (m!)^2 \\ \vdots \\ |\nu_m|\leq (m!)^2}} K_{m;\nu_1,\ldots,\nu_m}\, a\left(\frac{\nu_1}{m!}\beta_1 + \cdots + \frac{\nu_m}{m!}\beta_m; f\right) \\ &\qquad \times \exp\left[\mathrm{i}\left(\frac{\nu_1}{m!}\beta_1 + \cdots + \frac{\nu_m}{m!}\beta_m\right)t\right],[5] \end{aligned} \tag{26}$$

where

$$a(\lambda; f) = \boldsymbol{M}_t\{f(t)\exp(-\mathrm{i}\lambda t)\}.$$

Now we prove an important theorem.

Theorem. *For every almost periodic function f*

$$\lim_{m\to\infty} P_m(t; f) = f(t)$$

uniformly.

Proof. We consider the finite trigonometric polynomial

$$Q_N(t) = \sum_{n=1}^{N} b_n \exp(\mathrm{i}\lambda_n t) \quad (b_n \in X)$$

5 A finite basis $\{\beta_1, \beta_2, \ldots, \beta_r\}$ can be included in an infinite one $\{\beta_1, \beta_2, \ldots, \beta_r, \ldots, \beta_m, \ldots\}$. Here, if $|\nu_{r+1}| + |\nu_{r+2}| + \cdots + |\nu_m| \neq 0$, then

$$a\left(\frac{\nu}{m!}\beta_1 + \cdots + \frac{\nu_r}{m!}\beta_r + \cdots + \frac{\nu_m}{m!}\beta_m\right) = 0.$$

whose Fourier exponents $\lambda_n \in \tilde{B}$. Let m_0 and r_0 be chosen so that for $n = 1, 2, \ldots, N$

$$\lambda_n = \sum_{k=1}^{r_0} \nu_{k,n} \frac{\beta_k}{m_0!}, \tag{27}$$

and $|\nu_{k,n}| \leqslant (m_0!)^2$; it is not difficult to see that such a choice is possible. For $m \geqslant \max(m_0, r_0)$, from the representation (26) we obtain

$$P_m(t; Q_N) = \sum_{\substack{|\nu_1| \leqslant (m_0!)^2 \\ \vdots \\ |\nu_{r_0}| \leqslant (m_0!)^2}} \left(1 - \frac{|\nu_1|}{(m!)^2}\right) \cdots \left(1 - \frac{|\nu_{r_0}|}{(m!)^2}\right)$$

$$\times a\left(\frac{\nu}{m_0!}\beta_1 + \cdots + \frac{\nu_{r_0}}{m_0!}\beta_{r_0}; Q_N\right) \exp\left[\mathrm{i}\left(\frac{\nu_1}{m_0!}\beta_1 + \cdots + \frac{\nu_{r_0}}{m_0!}\beta_{r_0}\right)t\right].$$

Obviously,

$$\lim_{m\to\infty} P_m(t; Q_N) = Q_N(t) \tag{28}$$

uniformly. Now let $f(t)$ be an arbitrary almost periodic function. We consider for every $\varepsilon > 0$ the trigonometric polynomial $Q_\varepsilon(t; f)$ whose Fourier exponents belong to $\tilde{B}$ and for which

$$\sup_{t\in J} \|f(t) - Q_\varepsilon(t; f)\| \leqslant \varepsilon. \tag{29}$$

By the properties of the Fejer composite kernel, for every m we have

$$\sup_{t\in J} \|P_m(t; f) - P_m(t; Q_\varepsilon)\| = \sup_{t\in J} \|P_m(t; f - Q_\varepsilon)\|$$

$$\leqslant \sup_{t\in J} \|f(t) - Q_\varepsilon(t)\| \leqslant \varepsilon. \tag{30}$$

In view of (28), for a fixed ε, there is a sufficiently large $M = M(\varepsilon)$ such that

$$\sup_{t\in J} \|Q_\varepsilon(t; f) - P_m(t; Q_\varepsilon)\| \leqslant \varepsilon \tag{31}$$

for $m \geqslant M$. Then for every $t \in J$ and $m \geqslant M$, it follows from (29), (30), and (31) that

$$\|f(t) - P_m(t; f)\| \leqslant \|f(t) - Q_\varepsilon(t; f)\| + \|Q_\varepsilon(t; f) - P_m(t; Q_\varepsilon)\|$$

$$+ \|P_m(t; Q_\varepsilon) - P_m(t; f)\| \leqslant 3\varepsilon,$$

as we required.

5 Almost periodic functions with values in a Hilbert space; Parseval's relation

1. We are especially interested in the case when the space in which a function takes its values is a Hilbert space.

Let X be a complex Hilbert space, x, y be elements of X, (x, y) be a scalar product in X, and $\|x\| = (x, x)^{1/2}$ be the norm of $x \in X$.

Theorem. *For every almost periodic function*

$$f(t) \sim \sum_n a_n \exp(\mathrm{i}\lambda_n t) : J \to X,$$

Parseval's relation holds:

$$M_t\{(f(t), f(t))\} = \sum_{n=1}^{\infty} (a_n, a_n). \tag{32}$$

Proof. We take arbitrary elements $c_1, c_2, \ldots, c_n \in X$ and arbitrary numbers $\mu_1, \mu_2, \ldots, \mu_n \in J$ and consider the function $d = d(c_1, c_2, \ldots, c_n) = M_t\{\|f(t) - \sum_{k=1}^{n} c_k \exp(\mathrm{i}\mu_k t)\|^2\}$. We call $\sqrt{d}$ the *deviation* of the sum

$$s(t) = \sum_{k=1}^{n} c_k \exp(\mathrm{i}\mu_k t)$$

from the almost periodic function $f(t)$. We are going to find $\min_{c_i \in X} d(c_1, c_2, \ldots, c_n)$. In fact, it is easily obtained from an identity which is derived as follows:

$$d = M_t\left\{\left(f(t) - \sum_{k=1}^{n} c_k \exp(\mathrm{i}\mu_n t), f(t) - \sum_{l=1}^{n} c_l \exp(\mathrm{i}\mu_l t)\right)\right\}$$

$$= M_t\{(f(t), f(t))\} - \sum_{k=1}^{n} (c_k, M_t\{f(t) \exp(-\mathrm{i}\mu_k t)\})$$

$$- \sum_{l=1}^{n} (M_t\{f(t) \exp(-\mathrm{i}\mu_l t)\}, c_l)$$

$$+ \sum_{k=1}^{n} \sum_{l=1}^{n} (c_k, c_l) M_t\{\exp(\mathrm{i}\mu_k t) \exp(-\mathrm{i}\mu_l t)\}$$

$$= M_t\{(f(t), f(t))\} - \sum_{k=1}^{n} (c_k, a(\mu_k; f))$$

$$- \sum_{k=1}^{n} (a(\mu_k; f), c_k) + \sum_{k=1}^{n} (c_k, c_k)$$

$$+ \sum_{k=1}^{n} (a(\mu_k; f), a(\mu_k; f)) - \sum_{k=1}^{n} (a(\mu_k; f), a(\mu_k; f))$$

$$= \boldsymbol{M}_t\{(f(t), f(t))\} - \sum_{k=1}^{n} (a(\mu_k; f), a(\mu_k; f))$$

$$+ \sum_{k=1}^{n} (c_k - a(\mu_k; f), c_k - a(\mu_k; f)). \tag{33}$$

It is clear from this identity that min d is attained when and only when

$$c_k = a(\mu_k; f) = \boldsymbol{M}_t\{f(t) \exp(-i\mu_n t)\},$$

that is,

$$c_k = \begin{cases} 0 & \text{if } \mu_k \text{ does not coincide with any of the Fourier exponents of } f, \\ a(\mu_k; f) \neq 0 & \text{if } \mu_k \text{ coincides with one of the Fourier exponents of } f. \end{cases}$$

This property is called the *minimal* property of the Fourier coefficients. When we set $\mu_k = \lambda_k$ and $c_k = a(\lambda_k; f) = a_k$ in the identity (33) we obtain

$$d = \boldsymbol{M}_t\{(f(t), f(t))\} - \sum_{k=1}^{n} \|a_k\|^2. \tag{34}$$

Since the left-hand side of this last identity is non-negative, we obtain Bessel's inequality, namely:

$$\sum_{k=1}^{n} \|a_k\|^2 \leq \boldsymbol{M}_t\{(f(t), f(t))\}. \tag{35}$$

Here n is arbitrary so that we deduce, in particular, the convergence of $\sum_{k=1}^{\infty} \|a_k\|^2$. Now we are going to prove Parseval's relation (32); for this we take any $\varepsilon > 0$ and let

$$P_\varepsilon(t) = \sum_{k=1}^{n_\varepsilon} b_{k_\varepsilon} \exp(i\lambda_k t)$$

be a trigonometric polynomial for which[6]

$$\sup_{t \in J} \|f(t) - P_\varepsilon(t)\| \leq \varepsilon. \tag{36}$$

It follows from (36) and the minimal property of Fourier coefficients that

$$0 \leq d(a_1, a_2, \ldots, a_{n_\varepsilon}) \leq d(b_{1_\varepsilon}, b_{2_\varepsilon}, \ldots, b_{n_\varepsilon})$$
$$= \boldsymbol{M}_t\{\|f(t) - P_\varepsilon(t)\|^2\} \leq \varepsilon^2.$$

[6] We recall that the exponents of an approximating polynomial can be selected from the Fourier exponents of the function (see § 2, Property 3).

By combining this last inequality with the identity (34) we find that

$$0 \leqslant M_t\{\|f(t)\|^2\} - \sum_{k=1}^{n_\varepsilon} \|a_k\|^2 \leqslant \varepsilon^2,$$

and therefore since ε was chosen arbitrarily, we have proved Parseval's relation.

6 The almost periodic functions of Stepanov

1. Let $F: J \to X$ (X is a Banach space) be measurable in the sense of Lebesgue–Bochner. V. V. Stepanov suggested a generalisation of the concept of almost periodicity for this class of functions which is fully justified. Subsequently, Bochner pointed out that by using a very simple construction, a Stepanov function can be reduced to a Bohr function, which is vector valued even when the original is a scalar function; we reproduce this construction below. Let $\mathscr{L}^p(X)$ be the Banach space of measurable functions $\phi: \Delta = [0, 1] \to X$ with the norm

$$\left(\int_0^1 \|\phi(\eta)\|^p \,\mathrm{d}\eta\right)^{1/p} \qquad (p \geqslant 1).$$

Clearly, for every $t \in J$ the function $f(t+\eta)$ $(\eta \in \Delta)$ is a measurable function from Δ into X. We now state a definition of almost periodicity in the sense of Stepanov which takes account of the observation of Bochner mentioned above.

Definition. We say that a function $f(t): J \to X$ is *almost periodic in the sense of Stepanov* if $f(t) = \{f(t+\eta), \eta \in \Delta\}$ is almost periodic as a function $J \to \mathscr{L}^p(X)$.

More fully, $f(t)$ is almost periodic in the sense of Stepanov if for every $\varepsilon > 0$ there is a relatively dense set of numbers $\{\tau_\varepsilon\}$ satisfying the inequality[7]

$$\sup_{t \in J} \left(\int_0^1 \|f(t+\eta+\tau) - f(t+\eta)\|^p \,\mathrm{d}\eta\right)^{1/p} \leqslant \varepsilon.$$

The simplest properties of Stepanov almost periodic functions can be derived from the corresponding ones of Bohr almost periodic functions. For instance, since a Bohr almost periodic function is

[7] The continuity of $f(t+\eta)$ as a function from J into $\mathscr{L}^p(X)$ follows from the continuity in the mean of a function integrable in the sense of Lebesgue–Bochner.

bounded, we have

$$\sup_{t\in J}\int_0^1 \|f(t+\eta)\|^p \,d\eta<\infty. \tag{37}$$

Let $M^p(X)$ stand for the class of all measurable functions $f: J\to X$ satisfying (37), and $\mathring{M}^p(x)$ for the subclass of all Stepanov almost periodic functions. By using Bochner's criterion it can be proved that an element $f\in M^p(X)$ belongs to $\mathring{M}^p(X)$ if and only if the family of translates $f^t=\{f(t+s)\}$ is compact in $M^p(X)$.

We shall frequently find it convenient to use the following norm for the space $M^p(X)$ which is equivalent to (37):

$$\sup_{t\in J}\left(\frac{1}{l}\int_0^l \|f(t+\eta)\|^p \,d\eta\right\}^{1/p} \quad (l>0). \tag{38}$$

Lemma 4. *If a Stepanov almost periodic function is uniformly continuous as a function $J\to X$, then it is almost periodic.*

Proof. For any natural number n we set

$$f_n(t)=n\int_0^{1/n} f(t+\eta)\,d\eta.$$

Since $f(t)$ is uniformly continuous, it follows that $f_n(t)\to f(t)$ as $n\to\infty$ uniformly in t (see Chapter 1, § 1, Property 5). Therefore it is sufficient to prove that $f_n(t)$ is almost periodic for a fixed n. With this aim we use the norm (38) with $l=1/n$. For any $\varepsilon>0$ we can find a relatively dense set of numbers τ such that

$$\sup_{t\in J} n\int_0^{1/n} \|f(t+\tau+\eta)-f(t+\eta)\|^p \,d\eta\leqslant\varepsilon^p.$$

Then by using Hölder's inequality it follows that for any $t\in J$

$$\|f_n(t+\tau)-f_n(t)\|\leqslant n\int_0^{1/n} \|f(t+\tau+\eta)-f(t+\eta)\|\,d\eta$$

$$\leqslant nn^{-1/q}\left\{\int_0^{1/n} \|f(t+\tau+\eta)-f(t+\eta)\|^p \,d\eta\right\}^{1/p}\leqslant\varepsilon,$$

where $p^{-1}+q^{-1}=1$. This proves the lemma.

2. Now we are going to construct a Fourier series for a Stepanov almost periodic function. Since the function $\tilde f(t)=\{f(t+\eta),\ \eta\in\Delta\}$ is almost periodic in the sense of Bohr, it has the Fourier series

$$\tilde f(t)\sim\sum_{n=1}^{\infty} b_n \exp(i\lambda_n t),$$

where $b_n = b_n(\eta) \in \mathscr{L}^p(X)$, and

$$\lim_{T\to\infty} \int_0^1 \left\| b_n(\eta) - \frac{1}{2T} \int_{-T+\alpha}^{T+\alpha} f(t+\eta) \exp(-i\lambda_n t)\,dt \right\|^p d\eta = 0 \tag{39}$$

uniformly with respect to $\alpha \in J$. By making the substitution $t+\eta \to t$ and using (37) we find that

$$\lim_{T\to\infty} \int_0^1 \left\| b_n(\eta) \exp(-i\lambda_n \eta) - \frac{1}{2T} \int_{-T+\alpha}^{T+\alpha} f(t) \exp(-i\lambda_n t)\,dt \right\|^p d\eta = 0 \tag{40}$$

uniformly with respect to α. We use the notation:

$$b_n(\eta) \exp(-i\lambda_n \eta) = a_n(\eta),$$

$$\frac{1}{2T} \int_{-T+\alpha}^{T+\alpha} f(t) \exp(-i\lambda_n t)\,dt = c_n(T, \alpha).$$

From (40), by using Hölder's inequality we obtain that for every $h > 0$

$$\lim_{T\to\infty} c_n(T, \alpha) = \frac{1}{h} \int_h^{\eta+h} a_n(s)\,ds \quad (\eta \in [0, 1-h]) \tag{41}$$

uniformly with respect to α. In (41) we fix α and let $h \to 0$; then for almost all $\eta \in \Delta$

$$a_n(\eta) = \lim_{T\to\infty} c_n(T, \alpha) = a_n.$$

Therefore we can rewrite (41) as

$$\lim_{T\to\infty} c_n(T, \alpha) = a_n$$

uniformly with respect to α. Thus, we have proved that $b_n(\eta) \times \exp(-i\lambda_n \eta)$ is independent of η, that is, $b_n(\eta) \exp(-i\lambda_n \eta) = a_n \in X$ and

$$a_n = \lim_{T\to\infty} \frac{1}{2T} \int_{-T+\alpha}^{T+\alpha} f(t) \exp(-i\lambda_n t)\,dt,$$

uniformly with respect to α. The a_n are called the *Fourier coefficients* of the Stepanov almost periodic function $f(t)$. Thus,

$$\tilde{f}(t) = f(t+\eta) \sim \sum_{n=1}^{\infty} a_n \exp[i\lambda_n(t+\eta)].$$

By applying the Bochner summability method to $\tilde{f}(t)$ we obtain

$$\lim_{m\to\infty} \sup_{t\in J} \int_0^1 \left\| f(t+\eta) - \sum_{k=1}^{r_m} \mu_{mk} a_k \exp(i\lambda_k \eta) \exp(i\lambda_k t) \right\|^p d\eta = 0, \tag{42}$$

where the μ_{mk} are convergence factors that depend only on the spectrum $\{\lambda_n\}$. Obviously we can rewrite (42) in the form

$$\lim_{m\to\infty} \left\| f(t) - \sum_{k=1}^{r_m} \mu_{mk} a_k \exp(i\lambda_k t) \right\|_{\mathring{M}^p(X)} = 0,$$

which is an 'approximation theorem' for Stepanov almost periodic functions.

Comments and references to the literature

§ 1. The simple facts about Fourier–Stieltjes integrals given in this section are well documented in mathematical analysis. The reader can find a detailed exposition in Bochner's book [24].

§ 2. The first application of the Bochner–Khinchin theorem to the proof of basic theorems in the theory of (numerical) almost periodic functions is due to Bochner [24], who used this approach to derive Parseval's relation, from which it is comparatively simple to derive the approximation theorem (see, for example, Levitan [76]). Bochner's proof carries over easily to abstract almost periodic functions with values in a Hilbert space. Parseval's relation is meaningful for these functions (see § 5), but not for almost periodic functions with values in a Banach space (which is not a Hilbert space). The idea of the proof of the approximation theorem we have given goes back to one of the proofs of Bogolyubov (see Bogolyubov & Krylov [9]), who considered numerical almost periodic functions. It occurred to us that the proof of the approximation theorem could be considerably shortened by combining the basic idea of Bogolyubov's proof with the Bochner–Khinchin theorem. Amerio succeeded in carrying over Bogolyubov's proof to abstract almost periodic functions without using the Bochner–Khinchin theorem (see Amerio & Prouse [2]).

§ 4. Bochner–Fejer polynomials were first introduced by Bochner [25], who also extended them to abstract almost periodic functions [27].

§ 6. The Stepanov functions for numerical almost periodic functions were introduced by Stepanov in [104]. The special case corresponding to $p=2$ was also discussed by Wiener [33]. The reduction of Stepanov functions (numerical and abstract) to abstract almost periodic functions in the sense of Bohr was suggested by Bochner [27].

3 Arithmetic properties of almost periods

1 Kronecker's theorem

The following fundamental theorem of Kronecker about consistent solutions of a system of inequalities plays a significant role in what follows.

Kronecker's theorem. *Let $\lambda_1, \lambda_2, \ldots, \lambda_n$ and $\theta_1, \theta_2, \ldots, \theta_n$ be arbitrary real numbers. For the system of inequalities*

$$|\lambda_k t - \theta_k| < \delta \ (\text{mod } 2\pi) \quad (k = 1, 2, \ldots, n) \tag{1}$$

to have consistent real solutions for any arbitrarily small positive number δ, it is necessary and sufficient that every time the relation $l_1\lambda_1 + l_2\lambda_2 + \cdots + l_n\lambda_n = 0$ holds, where $l_1, l_2, \ldots, l_n$ are integers, we have the congruence

$$l_1\theta_1 + l_2\theta_2 + \cdots + l_n\theta_n \equiv 0 \ (\text{mod } 2\pi). \tag{2}$$

Proof. (*a*) *Necessity.* This is proved very simply. We assume that for every $\delta > 0$ there is a $t_\delta \in J$ such that (1) holds. We rewrite these inequalities in the form

$$-\delta < \lambda_k t_\delta - \theta_k - 2\pi n_k < \delta \quad (k = 1, 2, \ldots, n),$$

where the n_k are integers. By multiplying each of these inequalities by an l_k and adding them together we find that

$$-\delta \sum_{k=1}^{n} |l_k| < t_\delta \sum_{k=1}^{n} l_k\lambda_k - \sum_{k=1}^{n} l_k\theta_k - 2\pi \sum_{k=1}^{n} n_k l_k$$
$$< \delta \sum_{k=1}^{n} |l_k|.$$

If $\sum_{k=1}^{n} l_k\lambda_k = 0$, then since δ is arbitrary, from the last inequality we obtain

$$\sum_{k=1}^{n} l_k\theta_k \equiv 0 \pmod{2\pi},$$

as we required to prove.

(*b*) *Sufficiency.* The theorem will be proved if we can show that the upper bound of the absolute value of the function

$$f(t) = 1 + \exp[\mathrm{i}(\lambda_1 t - \theta_1)] + \cdots + \exp[\mathrm{i}(\lambda_n t - \theta_n)]$$

is equal to $n+1$, that is, to the value of the function of n independent complex variables

$$F(x_1, x_2, \ldots, x_n) = 1 + x_1 + x_2 + \cdots + x_n$$

at the point $(1, 1, \ldots, 1)$. Let p be an arbitrary positive integer; we set

$$\{f(t)\}^p = \sum_\nu \alpha_\nu \exp(\mathrm{i}\beta_\nu t), \tag{3}$$

$$\{F(x_1, x_2, \ldots, x_n)\}^p = \sum_{k_1, k_2, \ldots, k_n} a_{k_1, k_2, \ldots, k_n} x_1^{k_1} x_2^{k_2} \cdots x_n^{k_n}. \tag{4}$$

To obtain the expansion (3) from (4) we need to set $x_k = \exp[\mathrm{i}(\lambda_k t - \theta_k)]$ in (4), and then to reduce similar terms. If

$$k'_1\lambda_1 + k'_2\lambda_2 + \cdots + k'_n\lambda_n = k''_1\lambda_1 + k''_2\lambda_2 + \cdots + k''_n\lambda_n,$$

then by the condition in the theorem,

$$k'_1\theta_1 + k'_2\theta_2 + \cdots + k'_n\theta_n \equiv k''_1\theta_1 + k''_2\theta_2 + \cdots + k''_n\theta_n \pmod{2\pi}.$$

Therefore, the coefficients of similar terms have the same arguments. Since the absolute value of a sum of complex numbers with the same argument is equal to the sum of their absolute values,

$$\sum_\nu |\alpha_\nu| = \sum a_{k_1, \ldots, k_n} = \{F(1, 1, \ldots, 1)\}^p = (n+1)^p. \tag{5}$$

Now we assume that

$$\sup_{t \in J} |f(t)| = k < n+1. \tag{6}$$

Since

$$\alpha_\nu = \boldsymbol{M}_t\{(f(t))^p \exp(-i\beta_\nu t)\}, \tag{7}$$

from (6), (7) and an obvious estimate of the mean value, we obtain

$$|\alpha_\nu| \leq \boldsymbol{M}_t\{|f(t)|^p\} < k^p. \tag{8}$$

Since the number of terms in the expansion (4) is not greater than $(p+1)^n$,[1] it follows from (7) and (8) that

$$\sum_{\nu} |\alpha_\nu| \leq (p+1)^n k^p.$$

But for a sufficiently large p this inequality contradicts (5) because as $p \to \infty$

$$\frac{(p+1)^n k^p}{(n+1)^p} = (p+1)^n \left(\frac{k}{n+1}\right)^p \to 0.$$

This proves the sufficiency in Kronecker's theorem.

Remark. Since $f(t)$ is an almost periodic function, for every $\delta < \pi$ the set of solutions of the system of inequalities (1) is relatively dense. Hence it follows, in particular, that there are arbitrary large solutions of the system (1).

We end this section by mentioning two simple special cases of Kronecker's theorem that frequently occur in applications:

(*a*) all the $\theta_k = 0$ $(k = 1, 2, \ldots, n)$. In this case the conditions of Kronecker's theorem hold for any $\lambda_1, \lambda_2, \ldots, \lambda_n$. This means that the system of inequalities

$$|\lambda_k t| < \delta \pmod{2\pi} \quad (k = 1, 2, \ldots, n)$$

has solutions (and moreover arbitrarily large ones) for any real $\lambda_1, \lambda_2, \ldots, \lambda_n$ and an arbitrary $\delta > 0$;

(*b*) the numbers $\lambda_1, \lambda_2, \ldots, \lambda_n$ are linearly independent (relative to the integers), that is, the equation $l_1\lambda_1 + l_2\lambda_2 + \cdots + l_n\lambda_n = 0$ only

[1] It is not difficult to see this by using induction on n. For $n = 1$ and for all p the number $\Pi_{n;p}$ of terms in the expansion (4) is $(p+1)$, and consequently, the estimate holds. Next suppose that for all p and for a fixed $n = n_0$

$$\Pi_{n_0;p} \leq (p+1)^{n_0}.$$

We estimate $\Pi_{n_0+1;p}$. We have

$$(1 + x_1 + x_2 + \cdots + x_{n_0} + x_{n_0+1})^p = \sum_{k=0}^{p} \binom{p}{k} (1 + x_1 + \cdots + x_{n_0})^{p-k} x_{n_0+1}^k \tag{9}$$

Since we are interested in the *number* of terms (that is, the magnitude of the coefficients is immaterial), and since $\Pi_{n;p} < \Pi_{n;p'}$ for $p < p'$, it follows from (9) that

$$\begin{aligned} \Pi_{n_0+1;p} &= \sum_{k=0}^{p} \Pi_{n_0;k} \\ &\leq (p+1)\Pi_{n_0;p} \\ &\leq (p+1)(p+1)^{n_0} \\ &= (p+1)^{n_0+1}. \end{aligned}$$

holds if $l_1 = l_2 = \cdots = l_n = 0$. Under this assumption the condition of Kronecker's theorem holds for arbitrary $\theta_1, \theta_2, \ldots, \theta_n$. Therefore, in the case of linearly independent $\lambda_1, \lambda_2, \ldots, \lambda_n$, the system of inequalities (1) is solvable for any $\theta_1, \theta_2, \ldots, \theta_n$ and for any $\delta > 0$. This property can be expressed in a slightly different way: let $\xi_1, \xi_2, \ldots, \xi_n$ be arbitrary complex numbers with $|\xi_i| = 1$. Then the system of inequalities

$$|\exp(\mathrm{i}\lambda_k t) - \xi_k| < \delta \quad (k = 1, 2, \ldots, n)$$

has (arbitrarily large) solutions for any $\delta > 0$. In turn, we obtain from here the following result. Let $\xi = \{\xi_1, \ldots, \xi_n\}$ be an arbitrary point of the n-dimensional torus (that is, the product of n copies of the unit circle in the complex plane). We introduce a 'standard' shift on the torus:

$$\xi^t = \{\xi_1 \exp(\mathrm{i}\lambda_1 t), \xi_2 \exp(\mathrm{i}\lambda_2 t), \ldots, \xi_n \exp(\mathrm{i}\lambda_n t)\},$$

then for linearly independent $\lambda_1, \ldots, \lambda_n$ this shift is minimal, that is, every trajectory is everywhere dense on the torus.

2 The connection between the Fourier exponents of a function and its almost periods

1. As is well known, the period of a periodic function uniquely defines the set of its Fourier exponents: if $p > 0$ is the period, then all its Fourier exponents are multiples of $2\pi/p$. There is also a connection between the almost periods of an almost periodic function and its Fourier exponents, but clearly it is not as simple as in the periodic case. We establish this connection in the following two theorems.

Theorem 1. *For every $\delta > 0$ and for all natural numbers N there exists an $\varepsilon = \varepsilon(\delta; N) > 0$ such that every ε-almost period of an almost periodic function $f(t): J \to X$ satisfies the system of inequalities*

$$|\exp(\mathrm{i}\lambda_n \tau) - 1| \leq \delta \quad (n = 1, 2, \ldots, N). \tag{10}$$

Proof. *Let* $f(t) \sim \sum_n a_n \exp(\mathrm{i}\lambda_n t)$. Then

$$a_n = \boldsymbol{M}_t\{f(t)\exp(-\mathrm{i}\lambda_n t)\} \quad \Rightarrow \quad a_n \exp(\mathrm{i}\lambda_n \tau) = \boldsymbol{M}_t\{f(t+\tau)\exp(-\mathrm{i}\lambda_n t)\}.$$

Therefore,

$$\|a_n\| |\exp(\mathrm{i}\lambda_n \tau) - 1| = \|\boldsymbol{M}_t\{[f(t+\tau) - f(t)]\exp(-\mathrm{i}\lambda_n t)\}\|$$
$$\leq \boldsymbol{M}_t\{\|f(t+\tau) - f(t)\|\} \leq \varepsilon,$$

hence it follows that

$$|\exp(i\lambda_n\tau)-1|\leqslant\varepsilon/\|a_n\|.$$

If $k=\min_{1\leqslant n\leqslant N}\|a_n\|$, then

$$|\exp(i\lambda_n\tau)-1|\leqslant\varepsilon/k.$$

Therefore it is sufficient to set $\varepsilon=k\delta$.

Theorem 2 (*converse of the preceding theorem*). *For every $\varepsilon>0$ there exists a $\delta=\delta(\varepsilon)>0$ and a natural number $N=N(\varepsilon)$ such that every real number τ satisfying the system of inequalities* (10) *is an ε-almost period of an almost periodic function $f(t)$.*

Proof. In contrast to the proof of Theorem 1 which is completely straight-forward, that of Theorem 2 uses the deep approximation theorem.

For all $\varepsilon>0$ we consider the trigonometric polynomial

$$P_\varepsilon(t)=\sum_{n=1}^{N_\varepsilon} b_{n,\varepsilon}\exp(i\lambda_k t),$$

whose Fourier exponents are chosen from those of $f(t)$, and for which

$$\sup_{t\in J}\|f(t)-P_\varepsilon(t)\|\leqslant\tfrac{1}{3}\varepsilon. \tag{11}$$

We set $n=N_\varepsilon$ and $\delta=\varepsilon/3\sum_{n=1}^{N_\varepsilon}\|b_{n,\varepsilon}\|$. Suppose that τ satisfies the system of inequalities (1) with these n and δ. Then

$$\sup_{t\in J}\|P_\varepsilon(t+\tau)-P_\varepsilon(t)\|=\sum_{n=1}^{N_\varepsilon}\|b_{n,\varepsilon}\||\exp(i\lambda_n\tau)-1|$$

$$<\delta\sum_{n=1}^{N_\varepsilon}\|b_{n,\varepsilon}\|\leqslant\tfrac{1}{3}\varepsilon. \tag{12}$$

It follows from (11) and (12) that

$$\sup_{t\in J}\|f(t+\tau)-f(t)\|\leqslant\sup_{t\in J}\|f(t+\tau)-P_\varepsilon(t+\tau)\|$$

$$+\sup_{t\in J}\|P_\varepsilon(t+\tau)-P_\varepsilon(t)\|+\sup_{t\in J}\|P_\varepsilon(t)-f(t)\|\leqslant\varepsilon,$$

that is, τ is an ε-almost period of $f(t)$.[2]

[2] It follows from results in Chapter 4, §§2 and 3, that if we presuppose that the almost periods of an almost periodic function are solutions of a system of inequalities of the form (1), then the proof of the approximation theorem is comparatively simple. See also the theorem of Bogolyubov in Chapter 4.

2. Now we introduce a number of important new concepts which will be required later on.

Definition. A non-empty set of numbers is called a *module* if it is a group under the operation of addition.

For instance, the set of all real numbers, and the set of all real integers are modules. We shall only consider countable modules of real numbers.

Suppose that we are given a finite or countable set of real numbers $\alpha_1, \alpha_2, \ldots, \alpha_n, \ldots$ It is easy to form a module containing all the numbers of this set. In fact, it is enough to consider all the numbers of the form $m_1\alpha_1 + m_2\alpha_2 + \cdots + m_k\alpha_k$ with integers $m_1, m_2, \ldots, m_k$, where k is arbitrary and finite. Obviously, this set of numbers is the smallest module containing the set $\{\alpha_1, \alpha_2, \ldots, \alpha_n, \ldots\}$; we denote it by $\mathfrak{M}\{\alpha_1, \alpha_2, \ldots, \alpha_n, \ldots\}$.

Definition. A sequence of numbers $t_m \in J$ is called *f-increasing* if $f(t+t_m) \to f(t)$ uniformly, that is, t_m is an ε_m-almost period with $\varepsilon_m \to 0$.

Definition. A sequence of numbers $t_m \in J$ is called *f-normal* if the sequence $f(t+t_m)$ is fundamental in $C(X)$, that is,

$$f(t+t_m) \to \hat{f}(t) \tag{13}$$

uniformly.

It is useful to note that the uniform convergence (13) holds whenever there is convergence for each $t \in J$ (even on a set which is dense in J). This is easily deduced from the proof of the necessity of Bochner's criterion.

Now there arises the important problem of describing all the f-increasing and f-normal sequences.

Theorem 3. *For every almost periodic function $f: J \to X$ (X is a metric space) there exists a countable module $\mathfrak{M}_f$ with the following property: a sequence $\{t_m\}$ is f-increasing if and only if*

$$\exp(\mathrm{i}\lambda t_m) \to 1 \quad (\lambda \in \mathfrak{M}_f). \tag{14}$$

Proof. *If X* is a Banach space, then for $\mathfrak{M}_f$ we can take the smallest module containing all the Fourier–Bohr exponents of f. Then Theorem 3 follows immediately from Theorems 1 and 2.

In the general case we need to argue as follows. We choose on the compact metric set $\bar{\mathcal{R}}_f$ (recall that $\bar{\mathcal{R}}_f$ is the set of values of f) a countable sequence of scalar continuous functions $\phi_i(x)$ that separate any two points of this compact set.[3] Then it is clear that the convergence $f(t+t_m)\to f(t)$ $(t\in J)$ is equivalent to $\phi_i(f(t+t_m))\to \phi_i(f(t))$ $(t\in J, i=1,2,\ldots)$. But Theorem 3 is already proved for a scalar almost periodic function $\phi_i(f(t))$. Therefore, for $\mathfrak{M}_f$ we take the smallest module containing the Fourier–Bohr exponents of all the $\phi_i(f(t))$ $(i=1,2,\ldots)$, as is required.

Theorem 4. *For a sequence* $\{t_m\}$ *to be f-normal it is necessary and sufficient that the following limit exists:*

$$\lim_{m\to\infty} \exp(\mathrm{i}\lambda t_m)=\theta(\lambda) \quad (\lambda\in\mathfrak{M}_f). \tag{15}$$

Proof. From the equality

$$\sup_{t\in J} \rho(f(t+t_m), f(t+t_n)) = \sup_{t\in J} \rho(f(t+t_m-t_n), f(t))$$

we see that the f-normality of the sequence $\{t_m\}$ is equivalent to the sequence $\tau_{m,n}=t_m-t_n$ being f-increasing. On the other hand, from the equality

$$\exp(\mathrm{i}\lambda t_n)-\exp(\mathrm{i}\lambda t_m)=\exp(\mathrm{i}\lambda t_n)\{1-\exp(\mathrm{i}\lambda(t_m-t_n))\}$$

it follows that the limit (15) is equivalent to (14) for $\tau_{m,n}=t_m-t_n$. Thus, Theorem 4 is reduced to Theorem 3.

Now we make some observations regarding Theorem 4.

For an almost periodic function $f:J\to X$ we consider the family $\mathcal{H}=\mathcal{H}(f)$ (see Chapter 1, § 3). Each element $\hat{f}(s)\in\mathcal{H}$ is obtained as a limit $\hat{f}(s)=\lim_{m\to\infty} f(s+t_m)$. With each $\hat{f}(s)$ we associate the function

$$\begin{gathered} \theta(\lambda)=\lim_{m\to\infty}\exp(\mathrm{i}\lambda t_m) \quad (\lambda\in\mathfrak{M}_f), \\ \hat{f}(s)\to\theta(\lambda). \end{gathered} \tag{16}$$

Here, to the shift $\hat{f}^t(s)$ there corresponds the function $\theta(\lambda)\exp(i\lambda t)$, that is, a shift on $\mathcal{H}$ goes into the multiplication of $\theta(\lambda)$ by $\exp(i\lambda t)$.

This correspondence is one-to-one. For if we assume that two functions $\theta(\lambda)$ and $\theta^*(\lambda)$ correspond to $f(s)$, then there are two sequences $\{t_m\}$ and $\{t^*{}_m\}$ such that

$$\lim_{m\to\infty} f(s+t_m)=\lim_{m\to\infty} f(s+t^*{}_m)=\hat{f}(s),$$

$$\lim_{m\to\infty}\exp(\mathrm{i}\lambda t_m)=\theta(\lambda)\neq\theta^*(\lambda)=\lim_{m\to\infty}\exp(\mathrm{i}\lambda t^*{}_m).$$

[3] This means that $\phi_i(x)=\phi_i(y)$ $(i=1,2,\ldots)$ implies that $x=y$ (x, y are any two points of $\bar{\mathcal{R}}_f$).

It is obvious that the combined sequence $\{t_1, t^*_1, t_2, t^*_2, \ldots\}$ is f-normal, but it does not satisfy (15). Thus, the correspondence (16) is one-to-one. It can be studied in greater detail if we place additional arithmetical conditions on the module $\mathfrak{M}_f$. We restrict our discussion to the important case when $\mathfrak{M}_f$ has a finite integer basis $\beta_1, \beta_2, \ldots, \beta_n$. It is clear that in this case the convergence

$$\exp(i\lambda t_m) \to \theta(\lambda) \quad (\lambda \in \mathfrak{M}_f)$$

is equivalent to $\exp(i\beta_j t_m) \to \xi_j$ $(j = 1, 2, \ldots, n)$. Thus, we can associate with an $\hat{f} \in \mathscr{H}(f)$ a finite set of numbers $\xi = \{\xi_1, \xi_2, \ldots, \xi_n\}$. From Kronecker's theorem we obtain that in the case of a finite integer basis, a shift on $\mathscr{H}(f)$ is isomorphic to a standard shift on the n-dimensional torus.

3. From Theorems 3 and 4 it follows, in particular, that if two almost periodic functions f and g have the same modules ($\mathfrak{M}_f = \mathfrak{M}_g$), then they have one and the same set of increasing and of normal sequences. We are going to prove that if $\mathfrak{M}_f \neq \mathfrak{M}_g$, then the sets of f-increasing and g-increasing sequences are not the same.[4] Suppose that $\mathfrak{M}_f \neq \mathfrak{M}_g$, for example,

$$\mathfrak{M}_g \not\subset \mathfrak{M}_f. \tag{17}$$

We prove that if (17) holds, then there are arbitrarily exact almost periods of $f(t)$ that are not arbitrarily exact almost periods of $g(t)$. When (17) holds there are two possible cases:

(1) among the Fourier exponents of $g(t)$ there is a μ_{k_0} which is not representable as a linear combination of the Fourier exponents $\{\lambda_k\}$ of $f(t)$ with rational coefficients, that is, μ_{k_0} and $\{\lambda_1, \lambda_2, \ldots, \lambda_n, \ldots\}$ are linearly independent;

(2) there are no such Fourier exponents, that is, every Fourier exponent μ_k of $g(t)$ is representable in the form

$$l_k\mu_k = \sum_{j=1}^{N_k} \alpha_{k_j}\lambda_j, \tag{18}$$

where $l_k > 0$, the α_{k_j} are integers with no common divisor, and there exists at least one exponent μ_{k_0} for which $l_{k_0} > 1$ in every representation (18) (if $l_k = 1$, then $\mathfrak{M}_g \subset \mathfrak{M}_f$).

[4] Thus, for two almost periodic functions $f(t)$ and $g(t)$ to have the same increasing sequences, it is necessary and sufficient that $\mathfrak{M}_f = \mathfrak{M}_g$.

In the first case, by Kronecker's theorem the system of inequalities

$$\left.\begin{aligned}&|\lambda_k\tau|<\delta \pmod{2\pi} \quad (k=1,2,\ldots,N),\\ &|\mu_{k_0}\tau-\tfrac{1}{2}\pi|<\delta \pmod{2\pi}\end{aligned}\right\} \tag{19}$$

is solvable for any N and $\delta>0$. Therefore, if δ is sufficiently small and N sufficiently large, then τ is an ε-almost period of $f(t)$ (ε can be arbitrarily small). By the last inequality (19), τ cannot be an almost period of $g(t)$ (with arbitrarily high accuracy). We consider the second case. Let l'_{k_0} denote the smallest of the possible integers in the representation (18) for μ_{k_0}:

$$l'_{k_0}\mu_{k_0}=\sum_{j=1}^{N_{k_0}}\alpha_{k_0j}\lambda_j. \tag{18'}$$

By our condition, $l'_{k_0}>1$. We consider the system of inequalities

$$\begin{aligned}&|\lambda_k\tau|<\delta \pmod{2\pi} \quad (k=1,2,\ldots,N),\\ &|\mu_{k_0}\tau-2\pi/l'_{k_0}|<\delta \pmod{2\pi},\end{aligned} \tag{20}$$

where $\delta>0$ and N are arbitrary. By Kronecker's theorem, for the system (20) to be solvable it is necessary and sufficient that from every equation

$$l_{k_0}\mu_{k_0}=\sum_{j=1}^{N_k}\alpha_{kj}\lambda_j, \tag{21}$$

where $l_{k_0}>0$ and the α_{kj} are integers, follows the congruence

$$l_{k_0}\frac{2\pi}{l'_{k_0}}\equiv 0 \pmod{2\pi}.$$

Next we are going to prove that this congruence holds. Every l_{k_0} in the representation (21) is divisible by a minimal l'_{k_0}. For suppose that this is not so; then $l_{k_0}=m_{k_0}l'_{k_0}+h_{k_0}$, where $1\leq h_{k_0}<l'_{k_0}$. Therefore, the number $h_{k_0}\mu_{k_0}$ can be expressed as a finite linear combination of the λ_j with integer coefficients, which contradicts the fact that l'_{k_0} is the smallest of the possible integers in (18′). Now the proof is completed in exactly the same way as in the first case.

3 Limit-periodic functions

1. There is a deep connection between almost periodic functions and periodic functions of several variables (including a countable number). In this section we shall prove an important theorem of Bohr which asserts that every almost periodic function coincides with the values of a limit-periodic function (defined below) on the principal diagonal of the space.

Definition. A function $F(t_1, t_2, \ldots, t_n): R^n \to X$ is called a *limit-periodic* function of n variables $t_1, t_2, \ldots, t_n$ if it is the uniform limit (over the whole of R^n) of a sequence $F_k(t_1, t_2, \ldots, t_n)$ $(k = 1, 2, \ldots)$ of continuous periodic functions.

If all the $F_k(t_1, t_2, \ldots, t_n)$ have fixed periods in each of the variables, then it is obvious that the limit function is periodic with the same periods. We shall also require limit-periodic functions of a countable number of variables. Assume that we are given a sequence of continuous periodic functions $F_k(t_1, t_2, \ldots, t_{m_k})$, where $\overline{\lim}_{k\to\infty} m_k = \infty$. We define

$$\Phi_k(t_1, t_2, \ldots) = F_k(t_1, t_2, \ldots, t_{m_k}).$$

Thus, the functions Φ_k (of a countable number of variables) are constant with respect to $t_{m_k+1}, t_{m_k+2}, \ldots$ Now we define convergence. We say that a sequence $F_k(t_1, t_2, \ldots, t_{m_k})$ $(\overline{\lim}_{k\to\infty} m_k = \infty)$ converges to a function $F(t_1, t_2, \ldots)$ if the sequence $\Phi_k(t_1, t_2, \ldots)$ converges uniformly to $F(t_1, t_2, \ldots)$.

Definition. A function $F(t_1, t_2, \ldots)$ of a countable number of variables is called *limit-periodic* if there exists a sequence of continuous periodic functions $F_k(t_1, t_2, \ldots, t_{m_k})$ $(\overline{\lim}_{k\to\infty} m_k = \infty)$ which converges uniformly to $F(t_1, t_2, \ldots)$.

2. Let $F(t_1, t_2, \ldots, t_n)$ be a continuous periodic function with periods $q_1, q_2, \ldots, q_n$, and assume that $q_1^{-1}, q_2^{-1}, \ldots, q_n^{-1}$ are linearly independent. We consider the diagonal function $f(t) = F(t, t, \ldots, t)$. Subsequent results in this section are based essentially on the following theorem.

Theorem 5. *The set of values of the diagonal function $f(t)$ is everywhere dense in the set of values of $F(t_1, t_2, \ldots, t_n)$.*

Proof. We must prove that for all points $(t_1^{(0)}, \ldots, t_n^{(0)}) \in R^n$ and for every $\varepsilon > 0$ there is an $\xi \in J$ such that

$$\|F(t_1^{(0)}, t_2^{(0)}, \ldots, t_n^{(0)}) - f(\xi)\| < \varepsilon.$$

We choose $\delta = \delta(\varepsilon)$ so that

$$\|F(t'_1, t'_2, \ldots, t'_n) - F(t''_1, t''_2, \ldots, t''_n)\| < \varepsilon \tag{22}$$

for $|t'_i - t''_i| < \delta$ $(i = 1, 2, \ldots, n)$. Now we use Kronecker's theorem. By hypothesis the numbers $2\pi/q_i$ are linearly independent, and so there is an ξ and integers $k_1, k_2, \ldots, k_n$ such that

$$\left|\frac{2\pi}{q_i}\xi - \frac{2\pi}{q_i}t_i^{(0)} - 2\pi k_i\right| < \frac{2\pi\delta}{a} \quad (i = 1, 2, \ldots, n). \tag{23}$$

Here $a = \max_i |q_i|$. By multiplying both sides of (23) by $|q_i|/2\pi$ we obtain

$$|\xi - t_i^{(0)} - k_i q_i| < \delta \quad (i = 1, 2, \ldots, n), \tag{24}$$

and because the q_i are periods of $F(t_1, t_2, \ldots, t_n)$, it follows from (22) and (24) that

$$\|F(t_1^{(0)}, t_2^{(0)}, \ldots, t_n^{(0)}) - F(\xi, \xi, \ldots, \xi)\| < \varepsilon,$$

as we required to prove.

Corollary

$$\sup_{-\infty < t_i < \infty} \|F(t_1, \ldots, t_n)\| = \sup_{t \in J} \|f(t)\| \quad (i = 1, 2, \ldots, n).$$

Theorem 6. *Every almost periodic function $f(t)$ is the diagonal function of a limit-periodic function of a finite or countable number of variables.*

Proof. Let $f(t) \sim \sum_n a_n \exp(\mathrm{i}\lambda_n t) = \sum_n a_n \exp(\mathrm{i} \sum_{k=1}^{m_n} r_k \beta_k t)$ be an almost periodic function with a basis of Fourier exponents $\beta_1, \beta_2, \ldots, \beta_m, \ldots$ We denote by

$$s_k(t) = \sum_n b_n^{(k)} \exp[\mathrm{i}(r_1^{(n)}\beta_1 + r_2^{(n)}\beta_2 + \cdots + r_{\nu_k}^{(n)}\beta_{\nu_k})t]$$

a sequence of finite trigonometric polynomials which converges uniformly to $f(t)$. Every trigonometric polynomial is the diagonal function of a periodic function

$$F_k(t_1, t_2, \ldots, t_{\nu_k}) = \sum_k b_n^{(k)} \exp[\mathrm{i}(r_1^{(n)}\beta_1 t_1 + \cdots + r_{\nu_k}^{(n)}\beta_{\nu_k} t_{\nu_k})],$$

whose periods $q_1, q_2, \ldots, q_{\nu_k}$ are integer multiples of $2\pi/\beta_1$, $2\pi/\beta_2, \ldots, 2\pi/\beta_{\nu_k}$. The reciprocals of these periods are linearly independent. In exactly the same way, the periods of the difference

$$F_{k'}(t_1, t_2, \ldots, t_{\nu_{k'}}) - F_{k''}(t_1, t_2, \ldots, t_{\nu_{k''}})$$

are also linearly independent. Then by the corollary to Theorem 5 we have

$$\sup_{t_i \in J} \|F_{k'}(t_1, t_2, \ldots, t_{\nu_{k'}}) - F_{k''}(t_1, t_2, \ldots, t_{\nu_{k''}})\|$$
$$= \sup_{t \in J} \|s_{k'}(t) - s_{k''}(t)\| \quad (i = 1, 2, \ldots, \max(\nu_{k'}, \nu_{k''})).$$

Therefore, since the sequence $s_k(t)$ converges uniformly, the same holds for the sequence $F_k(t_1, t_2, \ldots, t_{\nu_k})$. Let $F(t_1, t_2, \ldots)$ denote the limit of the latter sequence, then $F(t_1, t_2, \ldots)$ is a limit-periodic function. By setting $t_1 = t_2 = \cdots = t$ we obtain

$$f(t) = \lim_{k\to\infty} s_k(t) = \lim_{k\to\infty} F_k(t, t, \ldots, t) = F(t, t, \ldots),$$

as we required to prove.

3. The nature of a limit-periodic function $F(t_1, t_2, \ldots)$ essentially depends on the kind of basis of Fourier exponents of $f(t)$ in $\mathfrak{M}_f$. For instance, if the basis in $\mathfrak{M}_f$ is integer, that is, every Fourier exponent of $f(t)$ is a linear combination of $\beta_1, \beta_2, \ldots$ with integer coefficients, then $F(t_1, t_2, \ldots)$ is periodic with periods $2\pi/\beta_1, 2\pi/\beta_2, \ldots$ If the basis is finite (but not necessarily integer), then $F(t_1, t_2, \ldots, t_n)$ is a limit-periodic function of a finite number of variables.

The case when a basis is both integer and finite is of special interest. Then $F(t_1, t_2, \ldots, t_n)$ is periodic in each variable. These almost periodic functions are called *conditionally periodic* (with periods $2\pi/\beta_1, 2\pi/\beta_2, \ldots, 2\pi/\beta_n$), or *quasiperiodic*.

Remark. Let $F(t_1, t_2, \ldots, t_n)$ be a continuous periodic function with periods $q_1, q_2, \ldots, q_n$; then the diagonal function

$$f(t) = F(t, t, \ldots, t)$$

is almost periodic. This assertion follows easily from Kronecker's theorem. Let us note that the basis of Fourier exponents of a diagonal function $f(t)$ is integer, but not necessarily with n elements. The function $f(t)$ can even turn out to be periodic, that is, with an integer basis of only one term. This holds when all the periods of $F(t_1, t_2, \ldots, t_n)$ are rational multiples of one and the same number.

4 Theorem of the argument for continuous numerical complex-valued almost periodic functions

1. In this section we assume that $f(t)$ is a numerical complex-valued almost periodic function satisfying the condition

$$\inf_{t\in J} |f(t)| \geq k > 0. \tag{25}$$

When this condition holds, the function $\phi(t) = f(t)/|f(t)|$ is defined and uniformly continuous on J and $|\phi(t)| = 1$. We set $\phi(t) = \exp(i\omega(t))$. If the value of $\omega(t)$ is chosen at some point t_0, then it can be naturally extended by continuity to other values of t. As a result, $\omega(t)$ is defined as a single-valued function with accuracy up to an integer multiple of 2π. It will be defined completely if as well as continuity we demand that $-\pi \leqslant \omega(0) < \pi$. The function $\omega(t)$ is called the *argument* of $f(t)$ and is denoted by $\arg f(t)$.

The next theorem has many important applications.

Theorem of the argument for an almost periodic function. *Suppose that a numerical complex-valued almost periodic function satisfies condition* (25). *Then*

$$\arg f(t) = ct + \psi(t), \tag{26}$$

where c is a constant and $\psi(t)$ *is a continuous almost periodic function. The number c and the Fourier exponents of* $\psi(t)$ *belong to the module* $\mathfrak{M}_f$ *of* $f(t)$.

Proof. First of all we are going to prove the theorem for a trigonometric polynomial

$$Q(t) = \sum_{n=1}^{N} a_n \exp(\mathrm{i}\lambda_n t).$$

We assume that the exponents of $Q(t)$ are contained in the module

$$\mathfrak{M} = \{h_1\mu_1 + \cdots + h_m\mu_m\} = \{\bar{h} \cdot \bar{\mu}\},$$

where $h_1, \ldots, h_m$ are integers, $\bar{h}$ is the vector with components $h_1, \ldots, h_m$, $\bar{\mu}$ is a constant vector with components $\mu_1, \ldots, \mu_m$, and $\bar{h} \cdot \bar{\mu}$ denotes the scalar product. Let

$$\lambda_1 = \bar{h}^{(1)} \cdot \bar{\mu}, \ldots, \lambda_N = \bar{h}^{(N)} \cdot \bar{\mu}.$$

In this notation

$$Q(t) = \sum_{n=1}^{N} a_n \exp(\mathrm{i}\bar{h}^{(n)} \cdot \bar{\mu} t).$$

We consider the function of m real variables $x_1, x_2, \ldots, x_m$

$$G(\bar{x}) = \sum_{n=1}^{N} a_n \exp(\mathrm{i}\bar{h}^{(n)} \cdot \bar{x}),$$

where $\bar{x} = (x_1, \ldots, x_m)$. By Theorem 5 of this chapter we have $|G(\bar{x})| \geqslant k$. We denote by $\arg G(\bar{x})$ an arbitrary continuous branch of the argument of $G(\bar{x})$. For every $l = 1, \ldots, m$ there exists an integer g_l, independent of $\bar{x}$ by the continuity of $\arg G(\bar{x})$, such that

$$\arg G(\ldots, x_l + 2\pi, \ldots) - \arg G(\ldots, x_l, \ldots) = g_l \cdot 2\pi. \tag{27}$$

Let $\bar{g}$ denote the vector $(g_1, \ldots, g_m)$. It follows from (27) that the function

$$\chi(\bar{x}) = \arg G(\bar{x}) - \bar{g} \cdot \bar{x} \tag{28}$$

is 2π-periodic in each variable. For,

$$\begin{aligned}\chi(\ldots, x_l+2\pi, \ldots)-\chi(\ldots, x_l, \ldots) \\ = \arg G(\ldots, x_l+2\pi, \ldots)-\arg G(\ldots, x_l, \ldots) \\ -\{g_1x_1+\cdots+g_l(x+2\pi)+\cdots+g_mx_m\} \\ +\{g_1x_1+\cdots+g_lx_l+\cdots+g_mx_m\}=0.\end{aligned}$$

When we set $x_1=\mu_1 t, \ldots, x_m=\mu_m t$ in (28) we obtain

$$\arg Q(t)=\arg G(\bar{\mu}t)=\bar{g}\cdot\bar{\mu}t+\chi(\bar{\mu}t)=ct+\psi(t),$$

where $c=\bar{g}\cdot\bar{\mu}\in\mathfrak{M}$, and $\psi(t)=\chi(\bar{\mu}t)$ is an almost-periodic function with Fourier exponents[5] belonging to $\mathfrak{M}$.

We can now prove the theorem in the general case by using the approximation theorem. Let $\varepsilon<\pi/6$ be any positive number and $Q_\varepsilon(t)$ be a trigonometric polynomial with Fourier exponents chosen from those of $f(t)$ (for example, a Bochner–Fejer polynomial) and satisfying the inequality

$$|f(t)-Q_\varepsilon(t)|\leqslant k\sin\varepsilon.$$

From this inequality and from (25) it follows that

$$|Q_\varepsilon(t)|\geqslant k(1-\sin\varepsilon)>k/2,$$

and so we can apply the theorem of the argument to $Q_\varepsilon(t)$, that is

$$\arg Q_\varepsilon(t)=c_\varepsilon t+\psi_\varepsilon(t). \tag{29}$$

Now we show that for chosen branches of $\arg f(t)$ and of $\arg Q_\varepsilon(t)$ we have the inequality

$$|\arg f(t)-\arg Q_\varepsilon(t)|\leqslant 2\varepsilon. \tag{30}$$

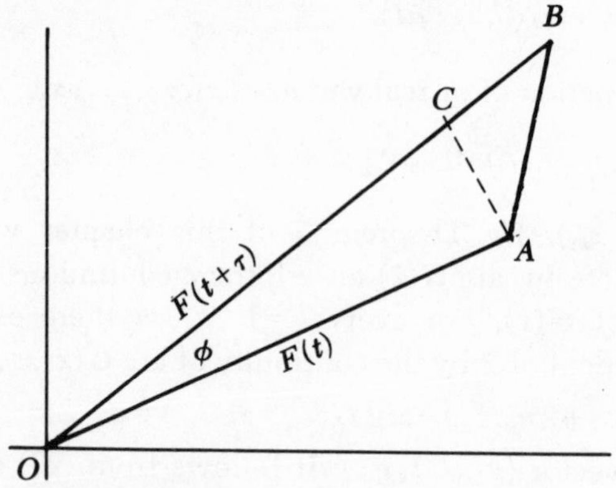

[5] The Fourier exponents of the periodic function $\chi(\bar{x})$ are integers. Therefore, the Fourier exponents of an almost periodic function $\psi(t)=\chi(\bar{\mu}t)$ have the form $k_1\mu_1+\cdots+k_m\mu_m$, where $k_1, \ldots, k_m$ are integers.

From the isosceles triangle OAC in this figure it is clear that

$$AC = 2OA \sin(|\phi|/2) < AB < k \sin \varepsilon. \tag{31}$$

Since $OA = k/2$, from (31) we obtain $\sin(|\phi|/2) < \sin \varepsilon$, from which it follows that $|\phi| < 2\varepsilon$. Therefore, for every $t \in J$ there is an integer n_t such that

$$|\arg f(t) - \arg Q_\varepsilon(t) - 2\pi n_t| < 2\varepsilon.$$

In view of the continuity of the arguments, the integer n_t does not depend on t. But $n_0 = 0$ for $t = 0$, and we obtain (30). From (29) and (30) it follows that

$$|\arg f(t) - c_\varepsilon t - \psi_\varepsilon(t)| < 2\varepsilon. \tag{32}$$

This inequality implies, in particular, that starting from some sufficiently small ε, c_ε does not depend on ε. For suppose that $\varepsilon' < \varepsilon < k \sin(\pi/6)$ and assume that $c_{\varepsilon'} \neq c_\varepsilon$.

Together with the inequality (32) the following inequality holds:

$$|\arg f(t) - c_{\varepsilon'} t - \psi_{\varepsilon'}(t)| < 2\varepsilon'. \tag{32'}$$

Then (32) and (32′) imply that

$$|(c_\varepsilon - c_{\varepsilon'})t - \psi_\varepsilon(t) + \psi_{\varepsilon'}(t)| < 2(\varepsilon + \varepsilon').$$

Since $\psi_\varepsilon(t)$ and $\psi_{\varepsilon'}(t)$ are almost periodic and so bounded functions, and since t can take arbitrarily large values, from the last inequality we obtain $c_\varepsilon - c_{\varepsilon'} = 0$, that is, $c_{\varepsilon'} = c_\varepsilon = c$. Thus, for $\varepsilon < k \sin(\pi/6)$, the inequality (32) can be written in the form

$$|\arg f(t) - ct - \psi_\varepsilon(t)| < 2\varepsilon. \tag{33}$$

Since $\psi_\varepsilon(t)$ is almost periodic for every $\varepsilon > 0$, it follows from (33) that $\arg f(t) - ct$ also is almost periodic. By denoting it by $\psi(t)$ we obtain the representation (26). The number c belongs to the module of $Q_\varepsilon(t)$ (for $\varepsilon < k \sin(\pi/6)$), and hence *a fortiori* to that of $f(t)$. Then the Fourier exponents of $\psi_\varepsilon(t)$ belong to the module of $Q_\varepsilon(t)$ for any $\varepsilon < k \sin(\pi/6)$, and so to that of $f(t)$. The same is true of the limit function $\psi(t)$ and the theorem of the argument is proved completely.

Comments and references to the literature

§ 1. The results of this section are classical. Kronecker's theorem (in the case of linearly independent $\lambda_1, \lambda_2, \ldots, \lambda_n$) has been refined in the following theorem of Kronecker–Weyl. Let $\lambda_1, \lambda_2, \ldots, \lambda_n$ be linearly independent real numbers, $\theta_1, \theta_2, \ldots, \theta_n$ be arbitrary real numbers, and $\delta_1, \delta_2, \ldots, \delta_n$ be arbitrary positive numbers. We denote by $f(x_1, \ldots, x_n)$ the characteristic function of the parallelepiped in R^n defined by the inequalities

$$\theta_k - \delta_k < x_k < \theta_k + \delta_k \quad (k = 1, 2, \ldots, n).$$

We extend $f(x_1, \ldots, x_n)$ to the whole space as a periodic function of period 2π in each variable.

Then we have uniformly in α

$$\lim_{(\beta-\alpha)\to\infty} \frac{1}{\beta-\alpha}\int_\alpha^\beta f(\gamma_1 t - \theta_1, \ldots, \gamma_n t - \theta_n)\, dt = \left(\frac{1}{\pi}\right)^n \delta_1 \cdots \delta_n.$$

A proof of the Kronecker–Weyl theorem is given in Levitan's book [76].

§ **2.** Theorems 1 and 2 are due to Bohr. The deep theorem that $\mathfrak{M}_f = \mathfrak{M}_g$ is a necessary and sufficient condition for the coincidence of f- and g-normal and f- and g-increasing sequences belongs to Favard [106].

§ **3.** The main ideas and results in § 3 (for the case of numerical almost periodic functions) are due to Bohr ([17], Part II). The original proofs of Bohr are far more involved than those given here, since in his development the connection between almost periodic functions and limit-periodic functions was established before the proof of the approximation theorem, and served mainly for the proof of the latter. The method we have presented belongs to Bochner [25]. It became possible only after simple proofs of the approximation theorem had been discovered. Obviously, there is no difficulty in carrying over the results to abstract almost periodic functions.

§ **4.** The theorem of the argument has its origin in mechanics (Lagrange's problem; see Jessen & Tornehave [60]). The version given by us was conjectured by A. Wintner; it was first proved by Bohr [20] but the proof we have presented belongs to Jessen [59]. For the generalisation to the case $f(t) \neq 0$, $\inf(f(t)) = 0$ see Levitan [78] and Gorin [37].

4 Generalisation of the uniqueness theorem (N-almost periodic functions)

1 Introductory remarks, definition and simplest properties of N-almost periodic functions

1. In this chapter we shall consider a class of generalised (continuous) almost periodic functions for which the uniqueness theorem holds in essentially the same form as for the classical almost periodic functions of Bohr. These generalised almost periodic functions have important applications, for instance, in the theory of differential equations with almost periodic coefficients. The concept of an ε, N-almost period is fundamental for this generalisation, and we begin by defining it.

Definition 1. Let ε and N be arbitrary positive numbers (ε is small and N is large). A number $\tau = \tau_{\varepsilon,N}$ is called an ε, N-almost period of a (continuous) function $f(t): J \to X$ if for all $|t| \leqslant N$

$$\|f(t \pm \tau) - f(t)\| \leqslant \varepsilon. \tag{1}$$

We could try to study the continuous functions for which there exists a relatively dense set of ε, N-almost periods for all $\varepsilon > 0$ and for all $N > 0$. But as Levin [71] has proved, this condition is not enough to distinguish a natural class of generalised almost periodic functions since the sum of two such functions may not have the property.

Thus, an additional condition must be imposed on the set of ε, N-almost periods; this condition is given in the following basic definition.

Definition 2. A function $f(t)$ continuous for all $t \in J$ and with values in a Banach space X, is called N-*almost periodic* if it satisfies the following two conditions:

(1) For all $\varepsilon > 0$ and for all $N > 0$ there exists a relatively dense set $E_{\varepsilon;N}$ of ε, N-almost periods of $f(t)$ (see Definition 1);

(2) for all $\varepsilon > 0$ and for all $N > 0$ there exists an $\eta = \eta(\varepsilon; N) > 0$ such that

$$E_{\eta;N} \pm E_{\eta;N} \subset E_{\varepsilon;N}.^{1}$$

In other words, if $\tau_i = \tau_i(\eta, N) \in E_{\eta;N}$ $(i = 1, 2)$, then their sum and difference also belong to $E_{\varepsilon,N}$, that is, they are ε, N-almost periods of $f(t)$. Obviously, every almost periodic function in the sense of Bohr is also N-almost periodic, but the converse is not true as will be seen from the example in subsection 4 (§ 1.4) below.

2. We now give another but equivalent definition of an N-almost periodic function, which often turns out to be more convenient to use.

Definition 3. A continuous function $f(t): J \to X$ is called *N-almost periodic* if there exists a fixed countable sequence of real numbers $\lambda_1, \lambda_2, \ldots, \lambda_k, \ldots$ such that for all $\varepsilon > 0$ and for all $N > 0$ there is a natural number $n = n(\varepsilon; N)$ and a positive number $\delta = \delta(\varepsilon; N)$ such that every $\tau \in J$ satisfying the system of inequalities

$$|\lambda_k \tau| \leq \delta \pmod{2\pi} \quad (k = 1, 2, \ldots, n) \tag{2}$$

also satisfies the inequality

$$\|f(t+\tau) - f(t)\| \leq \varepsilon \quad \text{for } |t| \leq N,$$

that is, it is an ε, N-almost period of $f(t)$.

Theorem 1. *Definitions* 2 *and* 3 *are equivalent.*

Proof. (*a*) Definition 3 implies Definition 2. For it follows from Kronecker's theorem that the set of solutions of the system (2) is relatively dense. Therefore, the first part of Definition 2 holds. To prove the second part, for given ε and N we define δ and n in accordance with Definition 3. Then we denote by $E_{\eta;N}$ the set of solutions of the system of inequalities

$$|\lambda_k \tau| \leq \delta/2 \pmod{2\pi} \quad (k = 1, 2, \ldots, n). \tag{2'}$$

Every $\tau \in J$ satisfying the system (2′) also satisfies the system (2).

[1] $E_{\eta;N}$ need not be the set of all η, N-almost periods, it can be only a part of it.

Therefore, $E_{\eta;N} \subset E_{\varepsilon;N}$, which means that every $\tau \in E_{\eta;N}$ is an η, N-almost period of $f(t)$ for some $\eta \leq \varepsilon$. Next it is obvious that

$$E_{\eta;N} \pm E_{\eta;N} \subset E_{\varepsilon;N},$$ [2]

which proves the second part of Definition 2.

(*b*) Definition 2 implies Definition 3. The proof of this assertion is based on the following theorem of Bogolyubov.

Theorem of Bogolyubov. *Let $\mathscr{E}$ be any relatively dense set of real numbers, δ any positive number, and $V = \{t \in J : |t| < \delta\}$ a neighbourhood of the origin. Then we can find a number $\eta > 0$ and a finite set of numbers $\omega_1, \omega_2, \ldots, \omega_n$ such that every τ satisfying the system*

$$|\exp(i\omega_k \tau) - 1| < \eta \quad (k = 1, 2, \ldots, n), \tag{3}$$

belongs to the set $\mathscr{E} - \mathscr{E} + \mathscr{E} - \mathscr{E} + V$.

Proof. We set $U = \{t \in J : |t| < \delta/4\}$. Then $U - U + U - U \subset V$. If l is a length in the definition of the relatively dense set $\mathscr{E}$ then this set has a non-empty intersection with any interval of length l. We partition J into intervals of the form $[4nl, 4(n+1)l]$, $n = 0, \pm 1, \pm 2, \ldots$. In each interval we fix an interval Δ_n of length $\delta/2$ that belongs to the set $\mathscr{E} + U$, and introduce the function

$$K = K_\delta(t) = \begin{cases} 1 & \text{if } t \in \Delta_n, \\ 0 & \text{if } t \notin \Delta_n. \end{cases}$$

It was shown in Chapter 2, § 2 that the following limit exists for some sequence $T_m \to \infty$:

$$\phi(x) = \phi_\delta(x) = \lim \frac{1}{2T_m} \int_{-T_m}^{T_m} K(t)K(t+x)\,dt$$

$$= M_t\{K(t)K(t+x)\}.$$ [3]

As is proved in § 2 of Chapter 2, $\phi(x)$ is positive-definite, and consequently has the representation

$$\phi(x) = a(x) + b(x)$$

$$= \sum_{k=0}^{\infty} A_k \exp(i\omega_k x) + \int_{-\infty}^{\infty} \exp(i\omega x)\,dS(\omega), \tag{4}$$

[2] We assume that $\delta < \pi$.

[3] $M\{g(t)\}$ denotes a limit point of the set

$$\frac{1}{2T} \int_{-T}^{T} g(t)\,dt \quad (T \to \infty).$$

where $A_k > 0$ and $S(\omega)$ is a continuous non-decreasing function. $\phi(x)$ has the property: if $\phi(x_0) > 0$, then $x_0 \in \mathscr{E} - \mathscr{E} + U - U \subset \mathscr{E} - \mathscr{E} + V$. For if $\phi(x_0) > 0$, then we have $K(t_0) > 0$ and $K(t_0 + x_0) > 0$ for some t_0, that is, $t_0 \in \mathscr{E} + U$ and $t_0 + x_0 \in \mathscr{E} + U$, and the required property follows immediately. Let the set of numbers $a_1, a_2, \ldots, a_q$ be a $(\delta/4)$-net of $[0, 4l]$. Then for every $x \in J$ at least one of the numbers $x + a_1, x + a_2, \ldots, x + a_q$ belongs to certain of the intervals Δ_k. Hence,

$$\sum_{i=1}^{q} K(x + a_i) \geqslant 1 \quad \text{for all } x \in J. \tag{5}$$

From this inequality it follows that

$$\sum_{i=1}^{q} \phi(x + a_i) = M_t\left\{K(t) \sum_{i=1}^{q} K(t + x + a_i)\right\} \geqslant M\{K(t)\}.$$

Since $M\{K(t)\} > 0$, by taking the mean of both sides of the last inequality we obtain $M\{\phi(x)\} > 0$. In the representation (4) we assume that $\omega_0 = 0$, that is, $A_0 = M\{\phi\} > 0$. Next we set

$$\psi(x) = \psi_\delta(x) = \lim_{T\to\infty} \frac{1}{2T} \int_{-T}^{T} \phi(x + t)\phi(t)\,\mathrm{d}t$$

$$= \lim_{T\to\infty} \frac{1}{2T} \int_{-T}^{T} \phi(x + t)\overline{\phi(t)}\,\mathrm{d}t.$$

From the properties of Fourier–Stieltjes integrals (see Chapter 2, § 1) we have

$$\psi(x) = \lim_{T\to\infty} \frac{1}{2T} \int_{-T}^{T} a(t + x)\overline{a(t)}\,\mathrm{d}t = \sum_{n=0}^{\infty} A_n^2 \exp(\mathrm{i}\omega_n x).$$

The function $\psi(x)$ has the property that from $\psi(x_0) > 0$ it follows that $x_o \in \mathscr{E} - \mathscr{E} + \mathscr{E} - \mathscr{E} + V$; this is proved in the same way as a similar property of $\phi(x)$. Then we choose n sufficiently large that

$$\sum_{k=n+1}^{\infty} A_k^2 \leqslant A_0^2/2.$$

Suppose that $\varepsilon < A_0^2/2$ is a positive number, and $\tau \in J$ satisfies the inequality

$$\sum_{k=1}^{n} A_k^2 |\exp(\mathrm{i}\omega_k \tau) - 1| < \varepsilon. \tag{6}$$

Then we have

$$\psi(\tau)=\sum_{k=0}^{\infty} A_k^2 \exp(i\omega_k\tau) \geq \frac{A_0^2}{2}+\sum_{k=1}^{n} A_k^2$$
$$+\sum_{k=1}^{n} A_k^2|\exp(i\omega_k\tau)-1| \geq \frac{A_0^2}{2}-\varepsilon>0,$$

that is, $\tau \in \mathscr{E}-\mathscr{E}+\mathscr{E}-\mathscr{E}+V$. This proves Bogolyubov's theorem, since for a sufficiently small η every solution of the inequalities (3) satisfies (6) (conversely, for a sufficiently small ε, every solution of (6) satisfies the system (3)).

Let $f(t)$ be an N-almost periodic function in the sense of Definition 2. Then for all $\varepsilon>0$ and for all $N>0$ there exists a relatively dense set $E_{\varepsilon;N}$ of ε, N-almost periods of $f(t)$. It follows from the continuity of $f(t)$ that there is a positive number $\delta=\delta(\varepsilon, N)<1$ such that

$$\|f(t+h)-f(t)\| \leq \varepsilon/2 \tag{7}$$

for $|h| \leq \delta$ and $|t| \leq N$. We use part 2 of Definition 2, selecting a positive number $\eta=\eta(\varepsilon; N)$ small enough that if $\tau_1, \tau_2, \tau_3, \tau_4 \in E_{\eta;N+1}$ then

$$\tau_1+\tau_2-\tau_3-\tau_4 \in E_{\varepsilon/2;N+1}. \tag{8}$$

Next we apply Bogolyubov's theorem taking the relatively dense set $\mathscr{E}$ to be the set $E_{\eta;N+1}$ and δ to be the number chosen above. Then from Bogolyubov's thereom it follows that there are real numbers $\omega_1, \omega_2, \ldots, \omega_n$ such that if $\tau \in J$ satisfies the system of inequalities

$$|\omega_k\tau| \leq \eta \pmod{2\pi} \quad (k=1, 2, \ldots, n), \tag{9}$$

then we can find numbers $\tau_1, \tau_2, \tau_3, \tau_4 \in E_{\eta;N+1}$ for which

$$|\tau-\tau^*| \leq \delta, \tag{10}$$

where $\tau^*=\tau_1+\tau_2-\tau_3-\tau_4$. It follows from (7), (8), and (10) that every number τ satisfying the system (9) is an ε, N-almost period for $f(t)$. To obtain the set $\lambda_1, \lambda_2, \ldots, \lambda_k, \ldots$ in Definition 3 we must take the union of a countable number of sets $\{\omega_1, \ldots, \omega_n\}$ for $(N=1, \varepsilon_1=1)$; $(N=2, \varepsilon_2=1/2); \ldots; (N=k, \varepsilon_k=1/k); \ldots$.

3. A sequence $\{t_m\} \subset J$ is called *returning* for a function $f: J \to K$ if

$$f(t+t_m) \to f(t)$$

uniformly on every finite interval. From Definition 3 we obtain the following criterion for N-almost periodicity.

Criterion for N-almost periodicity. *A function $f(t)$ is N-almost periodic if and only if there is a countable module $\mathfrak{M}=\mathfrak{M}_f$ such that every sequence $\{t_m\}$ satisfying the condition*

$$\exp(i\lambda t_m) \to 1 \quad (\lambda \in \mathfrak{M}) \tag{11}$$

is f-convergent.

From this criterion we obtain a natural method of constructing examples of N-almost periodic functions. Namely, let $\mathscr{H}$ be a compact metric space with a minimal two-sidedly stable dynamical system (see Chapter 1, § 3). We fix an $h_0 \in \mathscr{H}$ and consider the trajectory (a 'winding') $\{h_0^t\}=\{h_0(t)\}$ as an everywhere dense subset of $\mathscr{H}$. We define a continuous function $g(h)$ on $\{h_0(t)\}$, and claim that the function $f(t)=g(h_0^t)$ is N-almost periodic. To prove this we first note that $g(h)$ is uniformly continuous on the compact region $\{h_0(t): |t| \geqslant N\}$ (N is fixed), that is,

$$\|f(t)-f(s)\| < \varepsilon, \tag{12}$$

whenever $|t| \leqslant N$ and $\rho(h_0(t), h_0(s)) \leqslant \delta = \delta(N)$. Then since the trajectory $h_o(t)$ is almost periodic, the uniform 'returning' $h_0(t+t_m) \to h_0(t)$ is equivalent to condition (11) for some module $\mathfrak{M}$. From here and (12) we find that $\|f(t)-f(t+t_m)\| \leqslant \varepsilon$ $(|t| \leqslant N)$ whenever $\sup_{t \in J} \|h_0(t)-h_0(t+t_m)\| \leqslant \delta(N)$. This proves that $f(t)$ is N-almost periodic. We obtain especially intuitive examples by taking $\mathscr{H}$ to be a two-dimensional torus with the standard shift, and $g(h)$ to be a function without discontinuities on $\{h_0^t\}$.

4. To end this section we give a simple example of a (numerical) N-almost periodic function that is not almost periodic in the sense of Bohr. Let $\phi(t)=2+\cos t+\cos t\sqrt{2}$. It is easy to see that $\phi(t)>0$ for all $t \in J$, since if $\phi(t_0)=0$ then it would follow that $\cos t_0=-1$ and $\cos t_0\sqrt{2}=-1$, which is impossible because the numbers 1 and $\sqrt{2}$ are incommensurate. On the other hand, by Kronecker's theorem, for all $\delta>0$ there exists a $t_0 \in J$ satisfying the inequalities

$$|t_0-\pi|<\delta, \quad |t_0\sqrt{2}-\pi|<\delta \quad (\mathrm{mod}\, 2\pi).$$

Therefore, $\inf \phi(t)=0$; consequently, $\psi(t)=1/\phi(t)$ is unbounded and so cannot be almost periodic. We show that $\psi(t)$ is N-almost periodic. For this it is simpler to use Definition 3. For any $\delta>0$ there is an $\eta=\eta(\delta)$ such that every solution of the inequalities

$$|\tau| \leqslant \eta, \quad |\tau\sqrt{2}| \leqslant \eta \quad (\mathrm{mod}\, 2\pi)$$

is a δ-almost period for $\phi(t)$. We choose an arbitrary $N>0$ and set $k=\min_{|t|\leqslant N}\phi(t)$. Then for $|t|\leqslant N$ and $\delta<k$

$$|\psi(t+\tau)-\psi(t)|=\frac{|\phi(t+\tau)-\phi(t)|}{\phi(t)\phi(t+\tau)}\leqslant\frac{\delta}{k(k-\delta)}.$$

Therefore, if $\delta=\min(k/2, \varepsilon k^2/2)$, then $\delta/k(k-\delta)\leqslant\varepsilon$, and consequently τ is an ε, N-almost period for $\psi(t)$.

2 Fourier series, the approximation theorem, and the uniqueness theorem

1. For N-almost periodic functions (even bounded ones) the mean value

$$\boldsymbol{M}\{f(t)\}=\lim_{T\to\infty}\frac{1}{2T}\int_{-T}^{T}f(t)\,\mathrm{d}t$$

does not necessarily exist. Nevertheless, as we shall prove in this section, under certain natural conditions, with every N-almost periodic function we can associate a Fourier series. This correspondence is not one-to-one, because there are N-almost periodic functions with which one can associate a set of Fourier series. It is remarkable that in spite of this non-uniqueness, each of the Fourier series of an N-almost periodic function determines that function uniquely.

Let $f(t)$ be an N-almost periodic function, and $\lambda_1, \lambda_2, \ldots$ be a countable set of real numbers corresponding to $f(t)$ by Definition 3. We denote by $\mathfrak{M}_f=\mathfrak{M}(\lambda_1, \lambda_2, \ldots)=(\mu_1, \mu_2, \ldots)$ the smallest module for the set $\lambda_1, \lambda_2, \ldots, \lambda_n, \ldots$ (see Chapter 3, § 2). We show how to construct a Fourier series for an N-almost periodic function satisfying the condition

$$\overline{\lim_{T\to\infty}}\frac{1}{2T}\int_{-T}^{T}\|f(t)\|\,\mathrm{d}t<\infty. \tag{13}$$

We consider the second dual space X^{**}; it is a standard fact that the unit ball in X^{**} is weakly compact. For a fixed $\mu\in\mathfrak{M}_f$ we introduce the function

$$A_\mu(T)=\frac{1}{2T}\int_{-T}^{T}f(t)\exp(-\mathrm{i}\mu t)\,\mathrm{d}t \quad (T\geqslant 1),$$

and for every fixed $T\geqslant 1$ regard the set $\{A_\mu(T)\}$ ($\mu\in\mathfrak{M}_f$) as an element of the product of a countable number of copies of the space X^{**}. Then $\{A_\mu(T)\}$ has at least one limit point as $T\to\infty$; we denote it by $\{A_\mu\}=\{A_{\mu_1}, A_{\mu_2}, \ldots\}$, where $A_{\mu_k}=A_k\in X^{**}$. Thus, with each N-almost periodic function (satisfying (13)) we can associate a Fourier series

$$f(t) \sim \sum_k A_k \exp(\mathrm{i}\mu_k t). \tag{14}$$

We call the elements $A_k \in X^{**}$ the *Fourier coefficients* of $f(t)$. It is possible to produce simple conditions for the Fourier coefficients to belong to the original Banach space X, and by taking the (strong) limit with respect to some sequence $T_n \to \infty$, to obtain

$$A_k = \lim_{T_n \to \infty} \frac{1}{2T_n} \int_{-T_n}^{T_n} f(t) \exp(-\mathrm{i}\mu_k t)\, \mathrm{d}t \quad (\mu_k \in \mathfrak{M}_f).$$

For example, for this it is sufficient that $A_\mu(T)$ $(T \geq 1)$ has a compact trajectory in X for all $\mu \in \mathfrak{M}_f$.

2. For N-almost periodic functions the approximation theorem is valid in the following form.

Approximation theorem for N-almost periodic functions. *Let $f(t): J \to X$ be an N-almost periodic function satisfying condition* (13) *and with a Fourier series* (14). *For all $\varepsilon > 0$ and for all $N > 0$ there is a positive integer $Q = Q(\varepsilon, N)$ and real numbers $a_k = a_k(\varepsilon, N)$, $k = 1, 2, \ldots, Q$,* such that

$$\left\| f(t) - \sum_{k=1}^{Q} a_k A_k \exp(\mathrm{i}\mu_k t) \right\| \leq 2\varepsilon$$

for $|t| \leq N$.

Proof. By Definition 3 every $\tau \in J$ satisfying the system of inequalities (2) is an ε, N-almost period of $f(t)$. We construct on the interval $[-\pi, \pi]$ a smooth non-negative, even periodic function $\phi(t)$ with $\phi(0) > 0$ and $\phi(t) = 0$ for $|t| \geq \delta$. Then we define an almost periodic function $\psi(t)$ by

$$\psi(t) = \frac{1}{\alpha} \prod_{i=1}^{n} \phi(\lambda_i t),$$

where

$$\alpha = \lim_{T \to \infty} \frac{1}{2T} \int_{-T}^{T} \left(\prod_{i=1}^{n} \phi(\lambda_i t) \right) \mathrm{d}t.$$

It is obvious that $\psi(t)$ can be expanded as an absolutely convergent Fourier series:

$$\psi(t) = \sum_k a_k \exp(\mathrm{i}\mu_k t) \quad (\mu_k \in \mathfrak{M}_f). \tag{15}$$

Let $T_\alpha \to \infty$ be a generalised sequence used to determine the Fourier coefficients A_μ. We set

$$\begin{aligned} f_\varepsilon(t) &= \lim_{T_\alpha\to\infty} \frac{1}{2T_\alpha}\int_{-T_\alpha}^{T_\alpha} f(t+u)\psi(u)\,\mathrm{d}u \\ &= \lim_{T_\alpha\to\infty} \frac{1}{2T_\alpha}\int_{-T_\alpha}^{T_\alpha} f(u)\psi(t-u)\,\mathrm{d}u \\ &= \sum_k a_k A_k \exp(\mathrm{i}\mu_k t). \end{aligned} \tag{16}$$

Then we have

$$\begin{aligned} \|f(t)-f_\varepsilon(t)\| &= \left\| \lim_{T_\alpha\to\infty} \frac{1}{2T_\alpha}\int_{-T_\alpha}^{T_\alpha} f(t)\psi(u)\,\mathrm{d}u \right. \\ &\qquad \left. - \lim_{T_\alpha\to\infty} \frac{1}{2T_\alpha}\int_{-T_\alpha}^{T_\alpha} f(t+u)\psi(u)\,\mathrm{d}u \right\| \\ &\leqslant \overline{\lim_{T_\alpha\to\infty}} \frac{1}{2T_\alpha}\int_{-T_\alpha}^{T_\alpha} \|f(t)-f(t+u)\|\psi(u)\,\mathrm{d}u. \end{aligned} \tag{17}$$

If $\psi(u)\neq 0$, then $\|f(t)-f(t+u)\| \leqslant \varepsilon$ for $|t|\leqslant N$. Therefore, it follows from (17) that

$$\|f(t)-f_\varepsilon(t)\| \leqslant \varepsilon \lim_{T_\alpha\to\infty} \frac{1}{2T_\alpha}\int_{-T_\alpha}^{T_\alpha} \psi(u)\,\mathrm{d}u = \varepsilon. \tag{18}$$

Next, the absolute convergence of the Fourier series (14) implies that there is a Q such that

$$\sum_{k=Q+1}^{\infty} |a_k| \leqslant \varepsilon/\Gamma,$$

where $\Gamma = \overline{\lim}_{T\to\infty} (1/2T)\int_{-T}^{T}\|f(t)\|\,\mathrm{d}t$. Therefore

$$\left\| \sum_{k=Q+1}^{\infty} a_k A_k \exp(\mathrm{i}\mu_k t) \right\| \leqslant \varepsilon. \tag{19}$$

Then from (16), (18) and (19) it follows that

$$\left\| f(t) - \sum_{k=1}^{Q} a_k A_k \exp(\mathrm{i}\mu_k t) \right\| \leqslant 2\varepsilon,$$

as we required to prove.

3. Now let $f(t)$ and $g(t)$ be two N-almost periodic functions satisfying (13). By combining the sequences $\{\lambda_k\}$ in Definition 3 for $f(t)$ and $g(t)$, we can make the modules $\mathfrak{M}_f$ and $\mathfrak{M}_g$ coincide.[4]

[4] These modules do not have to be the smallest ones.

In constructing the Fourier series for $f(t)$ and $g(t)$ we are going to consider only common generalised sequences T_α.

Uniqueness theorem for N-almost periodic functions. *If $f(t)$ and $g(t)$ have the same Fourier series for at least one generalised sequence T_α, then they are identical.*

In fact, since in constructing Fourier series we are considering only common generalised sequences, and since by assumption the Fourier series for $f(t)$ and $g(t)$ are the same, it is clear from the construction of the approximating trigonometric polynomials that these polynomials for $f(t)$ and $g(t)$ are also the same. Therefore, $f(t)=g(t)$, as required.

Comments and references to the literature

§ 1. Definition 2 and the basic properties of N-almost periodic functions are due to Levitan ([73] and [74]). He was the first to show that the use of these functions leads to an interesting generalisation of a theorem of Favard about the solutions of ordinary differential equations with almost periodic coefficients (see Levitan [73] and [76]). Definition 3 belongs to Marchenko [86], while Bogolubov's theorem was first published in [6].

§ 2. Levin & Levitan [72] constructed an example of an N-almost periodic function with a non-unique Fourier series; the construction is given in Levitan's book [76]. Other examples of the non-uniqueness of a Fourier series, as well as conditions for uniqueness, are to be found in a deep paper of Levin [71]. Our construction of a Fourier series for an abstract N-almost periodic function is due to Zhikov.

The method of summing Bochner–Fejer series can be extended to N-almost periodic functions (for bounded N-almost periodic functions this was done by Marchenko [85], and by Levin [71] in the general case). We also mention that for an N-almost periodic function $f(t)$ satisfying the condition

$$\overline{\lim_{T\to\infty}}\frac{1}{2T}\int_{-T}^{T}\|f(t)\|^2\,dt<\infty,$$

one can easily extend the proof of the approximation theorem in Chapter 2, § 2.

A new class of generalised almost periodic functions was introduced by Bochner [28] in an interesting paper; they are called

Bochner almost-automorphisms. Almost automorphic functions have been studied in detail by Veech [34]. A numerical bounded uniformly continuous function $f(t)$ is called *almost automorphic* if for every sequence $\{t_m\}$ for which we have the local convergence

$$f(t+t_m)\xrightarrow{\text{loc}}\hat{f}(t),$$

the 'returning' also holds:

$$\hat{f}(t-t_m)\xrightarrow{\text{loc}}f(t).$$

Boles [10] and Reich [100] proved independently that an almost automorphic function is a bounded N-almost periodic function. These papers contain a more general definition of almost automorphic functions, and it is proved that the class of almost automorphic functions (in the sense of the slightly more general definition) coincides with the class of N-almost periodic functions. Baskakov [4] has proved that every scalar N-almost periodic function is the uniform limit of ratios of almost periodic functions.

5 Weakly almost periodic functions

1 Definition and elementary properties of weakly almost periodic functions

1. Let X be a Banach space and X^* the dual space. We denote the value of a continuous linear functional $x^* \in X^*$ at $x \in X$ by $x^*(x)$ or (x^*, x). From the linearity of elements of X^* it follows that for arbitrary $x^*_1, x^*_2, \ldots, x^*_n \in X^*$, arbitrary $x_1, x_2, \ldots, x_n \in X$, and any complex numbers $\alpha_1, \alpha_2, \ldots, \alpha_n; \beta_1, \beta_2, \ldots \beta_n$:

$$\left(\sum_{k=1}^{n} \alpha_k x^*_k, \sum_{j=1}^{n} \beta_j x_j\right) = \sum_{k=1}^{n} \sum_{j=1}^{n} \alpha_k \beta_j (x^*_k, x_j).$$

Let $\{x_n\}_{n=1}^{\infty}$ be a sequence of elements from X; $\{x_n\}$ is said to be *weakly convergent* (*weakly fundamental*) if the numerical sequence $x^*(x_n)$ is convergent (fundamental) for all $x^* \in X^*$. If in addition there exists an $x \in X$ such that

$$\lim_{n\to\infty} x^*(x_n) = x^*(x)$$

for all $x^* \in X^*$, then the sequence $\{x_n\}$ is said to *converge weakly* to x, and x is called the *weak limit* of $\{x_n\}$. We shall denote this by

$$\lim_{n\to\infty}{}^* x_n = x \quad \text{or} \quad x_n \xrightarrow{*} x.$$

It follows from the Hahn–Banach theorem that the weak limit is unique.

We call a Banach space X in which every weakly convergent sequence is weakly convergent to an $x \in X$, *weakly complete.*

Reflexive Banach spaces are weakly complete (since the dual spaces are weakly complete). In particular, every Hilbert space is weakly complete. On the other hand the space $C[0, 1]$ of numerical functions continuous on $[0, 1]$ is not weakly complete.

Definition. A function $f: J \to X$ is called *weakly almost periodic* if the numerical function $x^*(f(t))$ is a continuous almost periodic function for all $x^* \in X^*$.

This definition is similar to that of a weakly continuous or weakly measurable (summable) function.

2. We now consider certain elementary properties of weakly almost periodic functions.

Property 1. *$f(t)$ is an almost periodic function $\Rightarrow f$ is weakly almost periodic.*

This follows from the estimate

$$|x^*(f(t+\tau)) - x^*(f(t))| \leq \|x^*\| \|f(t+\tau) - f(t)\|.$$

Property 2. *$f(t)$ is a weakly almost periodic function $\Rightarrow \mathcal{R}_f$ is bounded and separable.*[1]

Proof For all $x^* \in X^*$ the numerical function $x^*(f(t))$ is almost periodic, and consequently bounded. Therefore, it follows from the Banach–Steinhaus theorem that

$$\sup_{t \in J} \|f(t)\| < \infty.$$

This proves that $\mathcal{R}_f$ is bounded. To prove separability we consider the countable set $\{f(r)\}$, where the r are rational points of J. For every $t_0 \in J$ there is a sequence $r_1, r_2, \ldots, r_k, \ldots (r_k \to t_0)$ of rational numbers such that

$$\lim_{k \to \infty}{}^* f(r_k) = f(t_0)$$

($f(t)$, being weakly almost periodic, is weakly continuous). Therefore, there is a sequence of linear combinations

$$y_n = \sum_{k=1}^{N_n} \alpha_{kn} x_k \quad (x_k = f(r_k), \alpha_{kn} \in C)$$

[1] Recall that $\mathcal{R}_f = \{x \in X : x = f(t), t \in J\}$.

which converges strongly to $f(t_0)$ as $n \to \infty$.[2] Without loss of generality we may assume that all the α_{kn} are rational. Hence the set of all y_n is countable and so $\mathcal{R}_f$ is separable.

Property 3. *Suppose that a sequence $\{f_n(t)\}$ of weakly almost periodic functions converges weakly to $f(t)$ uniformly in $t \in J$. Then $f(t)$ is weakly almost periodic.*

Proof. For all $x^* \in X^*$ we have

$$\lim_{n\to\infty} x^*(f_n(t)) = x^*(f(t)),$$

uniformly in $t \in J$. Therefore $x^*(f(t))$ is an almost periodic function and so $f(t)$ is weakly almost periodic.

Property 4. *Let f be a weakly almost periodic function and $\{s_n\}$ a sequence of real numbers for which*

$$\lim_{n\to\infty}{}^* f(t+s_n) = g(t) \quad \textit{for all } t \in J. \tag{1}$$

Then:

(i) *the convergence is uniform in $t \in J$;*
(ii) *if Ω_f denotes the convex hull of $\mathcal{R}_f$, then*

$$\bar{\Omega}_f = \bar{\Omega}_g; \tag{2}$$

(iii) $\sup_{t\in J} \|f(t)\| = \sup_{t\in J} \|g(t)\|$. (3)

Proof. (i) $x^*(f(t))$ is a numerical almost periodic function for all $x^* \in X^*$. Therefore, if $x^*(f(t+s_n))$ is convergent for all $t \in J$, then the convergence is uniform (see Chapter 1, § 2, the proof of the sufficiency in Bochner's thereom).

(ii) By definition, $\bar{\Omega}_f$ is the closure of the set Ω_f, that is, of the set of points

$$z = \sum_{j=1}^{p} \rho_j f(t_j), \quad t_1, t_2, \ldots, t_p \in J,$$

$$\sum_{j=1}^{p} \rho_j = 1, \quad \rho_j > 0.$$

We consider an arbitrary point of Ω_g:

$$y = \sum_{k=1}^{p} \rho_k g(t_k).$$

[2] See, for example, L. A. Lyusternik & V. I. Sobolev, *Elements of functional analysis*, 'Nauka', Moscow, 1965, p. 216 English translation, Frederick Ungar, New York, 1961, p. 123.

Obviously

$$y = \lim_{n\to\infty}{}^* \sum_{k=1}^{p} \rho_k f(t_k + s_n) = \lim_{n\to\infty}{}^* z_n,$$

where

$$z_n = \sum_{k=1}^{p} \rho_k f(t_k + s_n) \in \Omega_f.$$

Since the set $\bar{\Omega}_f$ is closed and convex, by a theorem of Mazur it is also weakly closed. Hence $y \in \bar{\Omega}_f$ and so

$$\bar{\Omega}_g \subseteq \bar{\Omega}_f.$$

Now we observe that for every fixed $x^* \in X^*$

$$\lim_{n\to\infty} x^*(f(t+s_n)) = x^*(g(t))$$

uniformly in $t \in J$. Therefore, for every $\varepsilon > 0$ there is a natural number n_ε (depending also on x^*) such that

$$\sup_{t\in J} |x^*(f(t+s_n)) - x^*(g(t))| \leqslant \varepsilon$$

for $n > n_\varepsilon$; hence it follows that

$$\sup_{t\in J} |x^*(f(t)) - x^*(g(t-s_n))| \leqslant \varepsilon,$$

that is

$$\lim_{n\to\infty} x^*(g(t-s_n)) = x^*(f(t))$$

uniformly in $t \in J$, or

$$\lim_{n\to\infty}{}^* g(t-s_n) = f(t). \tag{4}$$

Therefore,

$$\bar{\Omega}_f \subseteq \bar{\Omega}_g,$$

and so (ii) is proved.

(iii) From (i) and a standard result of functional analysis[3] it follows that

$$\|g(t)\| \leqslant \lim_{n\to\infty} \|f(t+s_m)\| \leqslant \sup_{t\in J} \|f(t)\|,$$

[3] See, fcr example, Lyusternik & Sobolev (footnote 2), p. 217; English translation p. 123.

and similarly from (4) we have

$$\|f(t)\| \leqslant \sup_{t\in J} \|g(t)\|,$$

which proves (3).

2 Harmonic analysis of weakly almost periodic functions

In this section we shall assume that the Banach space X is weakly complete (for instance, a reflexive space). We are going to prove certain additional properties of weakly almost periodic functions that are connected with their harmonic analysis.

Property 5. *$f(t)$ is a weakly almost periodic function $\Rightarrow$ the mean value*

$$a(\lambda) = \boldsymbol{M}^*\{f(t) \exp(-\mathrm{i}\lambda t)\}$$
$$= \lim_{T\to\infty}{}^* \frac{1}{2T}\int_{-T}^{T} f(t) \exp(-\mathrm{i}\lambda t)\, \mathrm{d}t \tag{5}$$

exists for all $\lambda \in J$. Moreover, the mean value

$$\lim_{T\to\infty}{}^* \frac{1}{2T}\int_{-T+s}^{T+s} f(t) \exp(-\mathrm{i}\lambda t)\, \mathrm{d}t \tag{6}$$

exists uniformly with respect to $s \in J$.

Proof. For all $\lambda \in J$ the function $f(t) \exp(-\mathrm{i}\lambda t)$, being weakly continuous, is Riemann integrable on every finite interval. Then the following mean value exists for all $x^* \in X^*$:

$$\lim_{T\to\infty} \frac{1}{2T}\int_{-T}^{T} x^*(f(t)) \exp(-\mathrm{i}\lambda t)\, \mathrm{d}t$$
$$= \lim_{T\to\infty}{}^* \left(\frac{1}{2T}\int_{-T}^{T} f(t) \exp(-\mathrm{i}\lambda t)\, \mathrm{d}t\right). \tag{7}$$

Since $x^* \in X^*$ is arbitrary and the space X is weakly complete, it follows from (7) that the weak limit

$$a(\lambda) = \lim_{T\to\infty}{}^* \frac{1}{2T}\int_{-T}^{T} f(t) \exp(-\mathrm{i}\lambda t)\, \mathrm{d}t$$
$$= \boldsymbol{M}^*\{f(t) \exp(-\mathrm{i}\lambda t)\}$$

exists. Then for all $x^* \in X^*$ the limit

$$\lim_{T\to\infty} \frac{1}{2T}\int_{-T+s}^{T+s} x^*(f(t)) \exp(-\mathrm{i}\lambda t)\, \mathrm{d}t = x^*(a(\lambda))$$

exists uniformly with respect to s. Therefore, the limit (6) exists uniformly with respect to s. By analogy with the earlier terminology $a(\lambda)$ is called the *Bohr transform* of $f(t)$.

Property 6. *$f(t)$ is a weakly almost periodic function* $\Rightarrow a(\lambda)=0$ *except for at most some sequence* $\{\lambda_n\}$.

Proof. By property 2 of § 1 $\mathcal{R}_f \subseteq X_0 \subseteq X$, where X_0 is a separable subspace of X. It is obvious that $a(\lambda) \in X_0$. Since X_0 is separable, there exists a determining sequence (with respect to X_0) of functionals $\{x^*_r\} \subset X^*$.[4] From a property of determining sequences of functionals it follows that

$$\|a(\lambda)\| = \sup_r |(x^*_r, a(\lambda))|. \tag{8}$$

For every fixed r, $(x^*_r, f(t))$ is a numerical almost periodic function, and so (for every fixed r) $(x^*_r, a(\lambda))=0$ except for some sequence $\{\lambda_k\}$. It follows from here and (8) that $a(\lambda)=0$ except for at most a sequence

$$\{\lambda_n\} = \bigcup_{k,r} \lambda_{k_r},$$

as we required to prove.

We set

$$a_n = a(\lambda_n), \tag{9}$$

and associate with $f(t)$ the Fourier expansion

$$f(t) \sim \sum_n a_n \exp(i\lambda_n t). \tag{10}$$

Let $B=\{\beta_n\}$ be a rational basis for the sequence $\{\lambda_n\}$ (see Chapter 2, § 4). We are going to extend the Bochner–Fejer summation procedure to weakly almost periodic functions. Let

$$P_m(t) = \sum_{k=1}^{N_m} \mu_{mk} a_k \exp(i\lambda_k t), \quad 0 \leq \mu_{mk} \leq 1,$$

be a Bochner–Fejer polynomial for $f(t)$ (see Chapter 2, § 4).

Property 7. *$f(t)$ is a weakly almost periodic function* $\Rightarrow \lim^* P_m(t) = f(t)$ *uniformly on J.*

Proof. First observe that B is also a basis for the Fourier exponents of the functions $(x^*, f(t))$ for all $x^* \in X^*$. In fact, it follows from (7)

[4] See, for example, Dunford & Schwartz [40].

that

$$a(\lambda; (x^*, f(t))) = \boldsymbol{M}\{x^*, f(t)) \exp(-\mathrm{i}\lambda t)\} = (x^*, a(\lambda)).$$

The last expression is zero for $\lambda \notin \{\lambda_n\}$. Furthermore,

$$(x^*, P_m(t)) = \sum_{k=1}^{N_m} \mu_{mk}(x^*, a(\lambda_k)) \exp(\mathrm{i}\lambda_k t),$$

that is, $(x^*, P_m(t))$ is a Bochner-Fejer polynomial constructed in terms of the basis B and the function $(x^*, f(t))$. Therefore, $\lim_{n\to\infty}(x^*, P_m(t)) = (x^*, f(t))$ uniformly on J as we required to prove.

Property 8. *$f(t)$ is a weakly almost periodic function and $a(\lambda) \equiv 0 \Rightarrow f(t) = 0$ (the uniqueness theorem for weakly almost periodic functions).*

Proof. If $a(\lambda) \equiv 0$, then $P_m(t) \equiv 0$ for all m. Therefore, $(x^*, f(t)) \equiv 0$ for all $x^* \in X^*$ and hence it follows that $f(t) \equiv 0$.

9. Bochner's criterion. *Let $f(t)$ be weakly continuous. For $f(t)$ to be weakly almost periodic it is necessary and sufficient that from each sequence $\{s_n\}$ we can extract a subsequence $\{s'_n\}$ such that $\{f(t+s'_n)\}$ is weakly convergent uniformly on J.*

Proof. The sufficiency of the condition is obvious. We shall prove the necessity. In the proof of Property 7 we remarked that for all $x^* \in X^*$ the Fourier exponents of the numerical almost periodic function $x^*f(t))$ are contained in a fixed countable set $\{\lambda_n\}$. It follows from this and Theorem 4 of Chapter 3 that it is enough to distinguish a sequence $\{s'_n\}$ satisfying the condition: the following limits exist for any $k(=1, 2, \ldots)$:

$$\lim_{n\to\infty} \exp(\mathrm{i}s'_n\lambda_k) = \theta_k;$$

this is clearly possible.

3 Criteria for almost periodicity

The following theorem gives a general criterion for almost periodicity.

Theorem 1. *For a bounded function $f: J \to X$ to be almost periodic it is necessary and sufficient that:*

(1) *For each x^* from a set D everywhere dense in X^* the scalar function $(x^*, f(t))$ is almost periodic;*

(2) *$f(t)$ is compact in the sense that the closure of the set of its values is compact.*

In particular, for a weakly almost periodic function to be almost periodic it is necessary and sufficient that it is compact.

Proof. The necessity of both conditions is obvious. We shall prove the sufficiency.

From the boundedness of $f(t)$ and condition (1) it follows that $f(t)$ is weakly almost periodic. In fact, for any $x^* \in X^*$ we can find a sequence of elements $x^*_n \in D$ such that $\|x^* - x^*_n\| \to 0$.

Then

$$|(x^*, f(t)) - (x^*_n, f(t))| \leqslant \|x^* - x^*_n\| \sup_{t \in J} \|f(t)\|,$$

from which it follows that $(x^*, f(t))$ is almost periodic.[5]

We proved earlier that the set $\mathcal{R}_f$ is separable. Therefore, without loss of generality we may assume that X itself is separable, and so is isomorphic to a subspace Y of the space of all functions continuous on the interval $[0, 1]$.[6]

We take a sequence of finite-dimensional linear operators $E_m : Y \to Y$ with the property

$$E_m y \xrightarrow[m \to \infty]{} y \quad (y \in Y) \tag{11}$$

(for this we can use any basis in the space of continuous functions).

The function $f_m(t) = E_m f(t)$ $(m = 1, 2, \ldots)$ has values in a finite-dimensional space, and so for it the concepts of almost periodicity and weak almost periodicity are equivalent; this is obtained from the equality

$$(y^*, f_m(t)) = (y, E_m f(t)) = (E_m{}^* y, f(t)),$$

where $E_m{}^*$ denotes the adjoint operator of E_m.

By (11), $f_m(t) \to f(t)$ for every $t \in J$, and in view of the next lemma this convergence is uniform with respect to $t \in J$.

Lemma 1. *The strong convergence of bounded linear operators is uniform on every compact set $K \subset Y$.*

Proof. Let $A_m : Y \to Y$ be a sequence of bounded operators for which $A_m y \to Ay$ for every $y \in Y$. Then by the Banach–Steinhaus theorem the norms $\|A_m\|$ are bounded by some number l.

[5] Thus, the boundedness of $f: J \to X$ and condition (1) are necessary and sufficient for the weak almost periodicity of the function.

[6] See L. A. Lyusternik & V. I. Sobolev (footnote 2), p. 256; English translation, p. 126.

Let $\{y_i\}$ $(i=1,2,\ldots,p)$ be a finite $(\varepsilon/4l)$-net for K, that is, for all $y \in K$ we can find a y_j $(j=1,\ldots,p)$ such that

$$\|y-y_j\| \leq \varepsilon/4l. \tag{12}$$

Furthermore, it is obvious that there is an $N=N(\varepsilon)$ such that

$$\|Ay_j - A_n y_j\| \leq \varepsilon/2 \tag{13}$$

for $n>N$ and for all $j=1,2,\ldots,p$. From (12) and (13) we obtain

$$\begin{aligned}\|Ay-A_ny\| &= \|(A-A_n)(y-y_j)+Ay_j-A_ny_j\| \\ &\leq \|(A-A_n)(y-y_j)\|+\|Ay_j-A_ny_j\| \\ &\leq 2l(\varepsilon/4l)+\varepsilon/2=\varepsilon\end{aligned}$$

for all $y \in K$. This proves the lemma, and also Theorem 1.

2. By using this general criterion for almost periodicity we obtain other special criteria which also have important applications. We begin with the following lemma.

Lemma 2. *Suppose that $f(t)$ is a weakly almost periodic function and that for a given sequence $\{s_n\}$*

$$\lim_{n\to\infty}{}^* f(t+s_n)=g(t) \quad \textit{uniformly}. \tag{14}$$

Then if $\|f(t)\|$ and $\|g(t)\|$ are almost periodic functions, there is a subsequence $\{s'_n\}\subset\{s_n\}$ such that

$$\lim_{n\to\infty} \|f(t+s'_n)\| = \|g(t)\| \quad \textit{uniformly}. \tag{15}$$

Proof. Since $\|f(t)\|$ is assumed to be almost periodic, there is a subsequence $\{s'_n\}\subset\{s_n\}$ such that

$$\lim_{n\to\infty} \|f(t+s'_n)\| = \phi(t) \tag{16}$$

uniformly, and $\phi(t)$ is almost periodic. It follows from (14) and (16) that [8]

$$\|g(t)\| \leq \lim_{n\to\infty} \|f(t+s'_n)\| = \phi(t) \tag{17}$$

From (14) it follows that

$$\lim_{n\to\infty}{}^* g(t-s_n)=f(t) \quad \text{uniformly}. \tag{18}$$

[8] If $x_0=\lim_{k\to\infty}{}^* x_n$, then $\|x_0\| \leq \underline{\lim}_{n\to\infty}\|x_n\|$ (see Lyusternik & Sobolev (footnote 2), p. 217, English translation, p. 123).

Since we have also assumed that $\|g(t)\|$ is almost periodic, then

$$\lim_{n\to\infty} \|g(t-s''_n)\| = \psi(t) \quad (\{s''_n\} \subset \{s'_n\}) \tag{19}$$

uniformly. Now (18) and (19) imply that

$$\|f(t)\| \leq \psi(t) \tag{20}$$

from which it follows that

$$\|f(t+s''_n)\| \leq \psi(t+s''_n),$$

and consequently, by (17) and (19)

$$\phi(t) = \lim_{n\to\infty} \|f(t+s''_n)\| \leq \lim_{n\to\infty} \psi(t+s''_n) = \|g(t)\|. \tag{21}$$

Now the conclusion (15) follows from (16), (18) and (21).

Let S_f denote the set of all sequences $s = \{s_n\}$ for which $f(t+s_n) \overset{*}{\rightharpoonup} f_s$ uniformly.

Theorem 2. *Suppose the following conditions are fulfilled:*

(*a*) *X is weakly complete;*

(*b*) *$x_n \overset{*}{\rightharpoonup} x$ and $\|x_n\| \to \|x\| \Rightarrow x_n \to x$;*

(*c*) *$\|f_s(t)\|$ is an almost periodic function for all $s \in S_f$.*

Then f is almost periodic.

Proof. It is enough to prove that the set $\mathcal{R}_f$ is compact. Let us assume otherwise. Then there is a $\rho > 0$ and a sequence $l = \{l_n\}$ such that

$$\|f(l_j) - f(l_k)\| \geq \rho \quad (j \neq k). \tag{22}$$

By Bochner's criterion we can extract from l a subsequence $s = \{s_n\} \in S_f$ such that

$$\lim_{n\to\infty}{}^* f(t+s_n) = f_s(t) \tag{23}$$

uniformly. Hence $f_s(t)$ is weakly almost periodic.

Since by hypothesis $\|f(t)\|$ is almost periodic, we can choose from s a subsequence $s' = \{s'_n\}$ for which $\|f(t+s'_n)\|$ converges uniformly. Then by using the preceding lemma we obtain

$$\lim_{n\to\infty} \|f(t+s'_n)\| = \|f_s(t)\|. \tag{24}$$

Next by using condition (*b*), from (23) and (24) we find that for all $t \in J$

$$\lim_{n\to\infty} \|f(t+s'_n) - f_s(t)\| = 0. \tag{25}$$

But relation (25) with $t=0$ contradicts (22), and so the theorem is proved.

There is an important class of Banach spaces for which conditions (*a*) and (*b*) of Theorem 2 are automatically fulfilled.

Definition. A Banach space X is called *uniformly convex* if for $0<\sigma\leq 2$ there exists a function $\omega(\sigma)$ with $0<\omega(\sigma)\leq 1$, such that for all $x_1, x_2 \in X$ satisfying

$$\|x_1\|\leq 1, \quad \|x_2\|\leq 1, \quad \|x_1-x_2\|\geq\sigma \tag{26}$$

we have

$$\|\tfrac{1}{2}(x_1+x_2)\|\leq 1-\omega(\sigma). \tag{27}$$

Condition (27) can also be stated as follows: for all $x_1, x_2 \in X$ satisfying

$$\|x_2-x_1\|\geq\sigma \max(\|x_1\|, \|x_2\|),$$

we have

$$\|\tfrac{1}{2}(x_1+x_2)\|\leq(1-\omega(\sigma)) \max(\|x_1\|, \|x_2\|). \tag{28}$$

Examples of uniformly convex spaces are the spaces l^p and $\mathscr{L}^p$ with $1<p<\infty$, and all Hilbert spaces. For a Hilbert space the assertion follows from the parallelogram equality

$$\|\tfrac{1}{2}(x_1+x_2)\|^2+\|\tfrac{1}{2}(x_1-x_2)\|^2=\tfrac{1}{2}(\|x_1\|^2+\|x_2\|^2).$$

Uniformly convex spaces are reflexive and consequently weakly complete. Condition (*b*) of Theorem 2 also holds. To see this we assume otherwise, then $\|x-x_{n_i}\|\geq\sigma>0$ for some subsequence $\{x_{n_i}\}\subset \{x_k\}$. Therefore

$$\|\tfrac{1}{2}(x+x_{n_i})\|\leq(1-\omega(\sigma)) \max(\|x\|, \|x_{n_i}\|).$$

By taking the weak limit we obtain the impossible inequality $\|x\|\leq (1-\omega(\sigma))\|x\|$.

The following result of Kadets [62] should also be noted: in any separable Banach space we can introduce an equivalent norm with the property (*b*).

3. It is possible to give quite simple conditions for almost periodicity in the case of a Hilbert space.

Let X be a separable Hilbert space and $\{e_n\}$ be an orthonormal basis in X. Then

$$f(t)=\sum_{k=1}^{\infty} e_k\phi_k(t), \quad \phi_k(t)=(f(t), e_k),$$

$$\|f(t)\|^2=\sum_{k=1}^{\infty} |\phi_k(t)|^2. \tag{29}$$

Theorem 3. *For a bounded function $f: J \to X$ to be almost periodic it is necessary and sufficient that:*

(1) *the scalar functions $\phi_k(t)$ are almost periodic;*

(2) *the series* (29) *is uniformly convergent.*

Proof. Suppose that $f(t)$ is almost periodic. Then since every functional on X has the form (x, h) (h is a fixed element of X), the functions $\phi_k(t)$ are almost periodic. Next we consider the projection operators $E_m: X \to X$ with $E_m x=\sum_{k=1}^{m} c_k e_k$ for $x=\sum_{k=1}^{\infty} c_k e_k$. By Lemma 1 the series $\sum_{k=1}^{\infty} e_k\phi_k(t)$ converges uniformly with respect to $t \in J$, that is, for any $\varepsilon>0$ we can find an $N=N(\varepsilon)$ such that for any $t \in J$

$$\left\|\sum_{k=N}^{\infty} e_k\phi_k(t)\right\|^2 = \sum_{k=N}^{\infty} |\phi_k(t)|^2 \leqslant \varepsilon.$$

Therefore, the series (29) is uniformly convergent.

Conversely, if the $\phi_k(t)$ are almost periodic and the series (29) is uniformly convergent, then the series $f(t)=\sum_{k=1}^{\infty} e_k\phi_k(t)$ is also uniformly convergent, that is, f is almost periodic.

4. We end this section by giving an example of a weakly but not strongly almost periodic function with values in a Hilbert space. Let $\{\phi_k(t)\}$ be a sequence of numerical almost periodic functions with the properties:

(1) the $\phi_k(t)$ are uniformly bounded, more precisely, $\sup_{t\in J}|\phi_k(t)|\leqslant 1$;

(2) the supports of $\phi_i(t)$ and $\phi_j(t)$ ($i \neq j$) are disjoint.

Then the function

$$f(t)=\sum_{k=1}^{\infty} e_k\phi_k(t)$$

is weakly but not strongly almost periodic. For by properties (1) and (2)

$$\sup_{t\in J} \sum_{k=1}^{\infty} |\phi_k(t)|^2 = 1.$$

Therefore $f(t)$ is a weakly almost periodic function (see footnote 5 on p. 71). Then, also by properties (1) and (2), for an arbitrary natural number N

$$\sup_{t\in J} \sum_{k=N}^{\infty} |\phi_k(t)|^2 = 1.$$

Therefore, the series $\sum_{k=1}^{\infty} |\phi_k(t)|^2$ cannot converge uniformly, and so it follows from Theorem 3 that $f(t)$ is not strongly almost periodic.

Comments and references to the literature

§§ 1 and 2. The concept of a weakly almost periodic function and all the properties mentioned in these sections are due to Amerio. It is interesting to note that for Banach spaces that are not weakly complete, it is not known whether every weakly almost periodic function has a countable spectrum.

§ 3. All the results in this section are also due to Amerio. The proof of the important Theorem 1 that we have given belongs to Zhikov. For other proofs of this theorem see Amerio & Prouse [2] and Levitan [77]. Theorem 3 is due to Amerio. We make an observation about the example at the end of § 3. In all it is simpler to construct a sequence of periodic functions with the same periods that satisfies conditions (1) and (2) of Theorem 3. Thus, there exists a weakly periodic function with values in a Hilbert space that is not strongly periodic.

6 A theorem concerning the integral and certain questions of harmonic analysis

1 The Bohl–Bohr–Amerio theorem

1. We are going to study the important question: if $f(t)$ is almost periodic, when is the indefinite integral $u(t)=\int_0^t f(t)\,dt$ almost periodic? In the periodic case there is a simple criterion for the periodicity of the integral, namely, the mean value of $f(t)$ is zero. But if $f(t)$ is almost periodic, when the mean value is equated to zero the spectrum can be condensed at zero, and so the indefinite integral can turn out to be unbounded.

We prove that for numerical functions, the almost periodicity of the indefinite integral follows from its boundedness. Clearly, it is enough to examine the case of real functions. Let $u(t)$ be a bounded solution of the differential equation $u'(t)=f(t)$, and let

$$m=\inf_{t\in J} u(t), \quad M=\sup_{t\in J} u(t).$$

We consider all possible sequences $\{t_m\}\subset J$ for which there is uniform convergence

$$f(t+t_m)\xrightarrow{J}\hat{f}(t), \tag{1}$$

and denote the set of limit functions $\hat{f}(t)$ by $\mathcal{H}(f)$. From the equality

$$u(t+t_m)=u(t_m)+\int_0^t f(s+t_m)\,ds \tag{2}$$

it is clear that the sequence $\{u(t+t_m)\}$ is compact in the sense of local convergence on J. Let $\hat{u}(t)$ be a limit point of this sequence; then obviously, $\hat{u}'(t)=\hat{f}(t)$. We prove

$$\hat{m}=\inf_{t\in J}\hat{u}(t)=m, \quad \hat{M}=\sup_{t\in J}\hat{u}(t)=M. \tag{3}$$

Since $\hat{u}(t)$ is a local limit of some subsequence of $\{u(t+t_m)\}$ we have $m \leqslant \hat{m} \leqslant \hat{M} \leqslant M$. To be specific we assume that $\hat{M} < M$.

The sequence $\{-t_m\}$ is 'returning', that is, $\hat{f}(t-t_m) \xrightarrow{J} f(t)$. Let $v(t)$ be a local limit point of $\{\hat{u}(t-t_m)\}$; then $v'(t)=f(t)$, that is $v(t)-u(t) \equiv C$. Then we have

$$\inf_{t \in J} (C+u) \geqslant \hat{m} \geqslant \inf_{t \in J} u,$$

$$\sup_{t \in J} (C+u) \leqslant \hat{M} \leqslant \sup_{t \in J} u,$$

which is impossible, and so the equalities (3) are proved. Clearly, if we add a non-zero constant to the solution $\hat{u}(t)$, then the equalities (3) are violated. Thus, with each $\hat{f}(t)$ from the class $\mathcal{H}(f)$ we can associate a unique preferred solution of the equation $u' = \hat{f}$. From here, in turn, we obtain the following observations.

Firstly, if $\{t_m\}$ satisfies (1), then the corresponding sequence $\{u(t+t_m)\}$ has a unique limit point in the sense of local convergence. In fact, assuming non-uniqueness we obtain two solutions of one and the same equation of the form $u' = \hat{f}$, and both solutions must be subject to (3).

Secondly, the convergence $u(t+t_m) \to \hat{u}(t)$ is actually uniform on the whole line. For if this were not the case, then we could find a sequence $\{s_m\} \subset J$ such that

$$|u(s_m+t_m) - \hat{u}(s_m)| \geqslant \alpha > 0. \tag{4}$$

Then by going over to subsequences, if necessary, we can assume that $\{f(s_m+t_m+t)\}$ and $\{\hat{f}(s_m+t)\}$ are fundamental in the sense of uniform convergence. Here it is important that the limits of these sequences must coincide. But then by (4) the local limits of $\{u(s_m+t_m+t)\}$ and $\{\hat{u}(s_m+t)\}$ are different, which is impossible.

The general case of abstract functions turns out to be more complicated. To appreciate this it is enough to observe that the boundedness of the indefinite integral does not by any means imply almost periodicity (an illustrative example is given in § 2); in fact, we can only assert that boundedness implies weak almost periodicity. For this, let $y \in X^*$, then

$$(y, u(t)) = \int_0^t (y, f)\, ds$$

and use the result we have proved for numerical functions.

Theorem 1 in Chapter 5 shows that almost periodicity is a consequence of weak almost periodicity and compactness, and so we need to study the question of the compactness of the indefinite integral. It turns out that the problem of compactness is closely connected with geometrical properties of the domain of values, that is, the Banach space X. The most general solution of this problem will be presented in § 2, but for the time being we restrict ourselves to the important special case of a uniformly convex space.

Let X be a uniformly convex space and $f: J \to X$ be an almost periodic function with a bounded indefinite integral $u(t) = \int_0^t f(s)\,\mathrm{d}s$. We set $M = \sup_{t \in J} \|f(t)\|$.

From preceding results for numerical functions we obtain that if a sequence $\{t_m\}$ satisfies condition (1), then the sequence $(y, u(t+t_m))$ $(y \in X^*)$ is fundamental in the sense of uniform convergence. The space X is weakly complete, therefore $(y, u(t+t_m)) \xrightarrow{J} (y, \hat{u}(t))$. Since taking a weak limit does not increase norms, we have

$$\hat{M} = \sup_{t \in J} \|\hat{u}(t)\| \leqslant \sup_{t \in J} \|u(t+t_m)\|$$

$$= \sup_{t \in J} \|u(t)\| = M.$$

But the sequence $\{-t_m\}$ is 'returning', that is, we have the convergence $(y, \hat{u}(t-t_m)) \xrightarrow{J} (y, u(t))$. Therefore, $\hat{M} = M$.

We recall that a space X is called *uniformly convex* if from the inequality

$$\|x_1 - x_2\| \geqslant \rho \max\{\|x_1\|, \|x_2\|\}$$

it follows that

$$\|(x_1+x_2)/2\| \leqslant (1-\phi(\rho)) \max\{\|x_1\|, \|x_2\|\} \quad (\phi(\rho) > 0).$$

Now we are going to show that $u(t)$ is compact. If we assume otherwise, then we would have a sequence $\{t_k\} \in J$ such that

$$\|u(t_m) - u(t_n)\| \geqslant \rho M > 0 \quad (m \neq n, \rho > 0).$$

Therefore, by taking subsequences if necessary, we can assume the convergence:

$$f(t+t_k) \xrightarrow{J} \hat{f}(t), \quad (y, u(t+t_k)) \xrightarrow{J} (y, \hat{u}(t)).$$

Furthermore, for any fixed $t \in J$ we have

$$\|u(t+t_m)-u(t+t_n)\| = \left\|u(t_m)-u(t_n)+\int_0^t |f(s+t_m)-f(s+t_n)|\,\mathrm{d}s\right\|$$

$$\geqslant \rho M - \int_0^t \|f(s+t_m)-f(s+t_n)\|\,\mathrm{d}s = \rho_{m,n}(t)M.$$

Since $\rho_{m,n}(t) \to \rho$ as $m, n \to \infty$ $(m \neq n)$, for any $t \in J$ we have

$$\|\hat{u}(t)\| \leqslant \lim_{m \neq n \to \infty} \frac{\|u(t+t_m)+u(t+t_n)\|}{2}$$

$$\leqslant \lim_{m \neq n \to \infty} (1-\phi(\rho_{m,n}(t))M \leqslant (1-\phi(\rho))M < M.$$

This contradicts the equality $\hat{M} = M$ that we established earlier, and so the compactness of $u(t)$ is proved.

It is not difficult to check that our arguments in both numerical and abstract cases are only slightly changed if $f(t)$ is almost periodic in the sense of Stepanov rather than in the sense of Bohr, that is, it is an element of $\mathring{M}^1(X)$. Thus, the following result holds

Theorem 1 (*The Bohl–Bohr–Amerio theorem*). *If X is a uniformly convex space, then from the boundedness of an indefinite integral of an almost periodic function $f(t)$ (and even of a Stepanov almost periodic function) follows its almost periodicity.*

2. There are more general classes of functions for which the theorem on the integral holds. They are characterised by one or another property of being 'returning'.

Let $f: J \to X$ be a continuous function. A sequence of real numbers $t_m \to \infty$ is called *returning* (more precisely, *f-returning*) if

$$f(t+t_m) \xrightarrow{\text{loc}} f(t).$$

Definition 1. A function $f(t)$ is *stable in the sense of Poisson* if it has at least one returning sequence.

Definition 2. A continuous function $f: J \to X$ is called *recurrent* if for any $\varepsilon, N > 0$ the set

$$L(\varepsilon, N, f) = \{\tau \in J : \sup_{|t| \leqslant N} \|f(\tau+t)-f(t)\| < \varepsilon\}$$

is relatively dense on the real line.

Obviously, the class of almost periodic functions belongs to the class of recurrent functions, which belongs to the class of functions stable in the sense of Poisson.

Proposition 1. *If the indefinite integral of a Poisson stable (respectively, recurrent) function is compact, then it is Poisson stable (respectively, recurrent).*

Proof. First we consider the Poisson stable case.

It is clear from (2) that the sequence $\{u(t+t_m)\}$ is compact in the sense of local convergence. Let $v(t)$ be a limit point of this sequence, that is, $v(t)=\lim u(t+s_m)$, where $\{s_m\}$ is a subsequence of $\{t_m\}$. Clearly, $v(t)=u(t)+c$, where c is a constant element from X.

We denote by Tg the local limit of the subsequence $\{g(t+s_m)\}$, assuming that it exists. Since $Tu=u+c$ we have

$$\begin{aligned} &T^2u=u+2c, \quad T^3u=u+3c, \ldots, \\ &T^nu=u+nc. \end{aligned} \tag{5}$$

Because $\sup_{t\in J}\|Tg\|\leqslant \sup_{t\in J}\|g\|$, for a sufficiently large n the equality (5) is impossible for $c\neq 0$. Thus we obtain $v(t)\equiv u(t)$, which clearly means that the sequence $\{t_m\}$ is u-returning.

The following property holds for recurrent functions: for any $\varepsilon, N>0$ we can find a $\delta=\delta(\varepsilon, N)$ such that

$$L(\delta, N, f)\subset L(\varepsilon, N, u).$$

It is not difficult to prove this property by contradiction, using the conservation of returning sequences that has been proved already.

In the next section the requirement that the integral is compact will be weakened significantly.

2. Further theorems concerning the integral

1. Let c_0 denote the Banach space of numerical sequences $\kappa=\{\xi_i\}$ $(i=1, 2, \ldots)$ that converge to 0 with the norm $\|\kappa\|=\max\{|\xi_i|\}$. We consider the function $f: J\to c_0$ defined by $f(t)=\{(1/n)\cos(t/n)\}$. It is easily seen that $f(t)$ is weakly almost periodic and compact, and that the indefinite integral $u(t)=\int_0^t f(s)\,ds$ has the form $\{\sin(t/n)\}$. Since every functional on c_0 has the form $\sum \xi_i\eta_i$, where $\sum|\eta_i|<\infty$, $u(t)$ is weakly almost periodic. However, $u(t)$ is not compact. For consider the sequence of functionals $\Phi_n(\kappa)=\xi_n$. If the set $\{u(t), t\in J\}$ were compact, then the convergence $\Phi_n(\kappa)\to 0$ would be uniform on it. But $\Phi_n(u(t))=\sin(t/n)\to 0$ non-uniformly with respect to $t\in J$.

Thus, c_0 contains an almost periodic function whose indefinite integral is bounded but not compact. It might seem from the simplicity of the preceding construction that similar counterexamples are possible for other Banach spaces. Fortunately, this expectation is not justified. The remarkable fact is that if a space X has no subspace isomorphic to c_0, then no similar counterexample is possible.

We say that a space X_n *does not contain* c_0 if it has no subspace isomorphic to c_0. A reflexive or even a weakly complete space does not contain c_0. For what follows we need an explicit characterisation of spaces that do not contain c_0.

Suppose that we have a series $\sum_{n=1}^{\infty} e_n$ $(e_n \in X)$; we denote by Δ the set of all partial sums obtained from any permutation of the terms of this series. We call the series *unconditionally bounded* if the set Δ is bounded.

If in c_0 we take the sequence of unit vectors $e_1 = \{1, 0, 0, \ldots\}$, $e_2 = \{0, 1, 0, \ldots\}$, $\ldots$, then we obtain a divergent unconditionally bounded series $\sum e_n$. In a certain sense the converse assertion also holds.

Proposition 2. *If there exists a divergent unconditionally bounded series, then X contains* c_0.

Proof. Let a be any element in Δ, that is, an arbitrary partial sum. We set

$$\lambda(a) = \sup_{|\alpha_i| \leq 1} \|\sum \alpha_i e_i\|,$$

where the α_i are real numbers and the summation is taken over all elements e_i in a. From Abel's identity we easily find that

$$\sup_{a \in \Delta} \lambda(a) \overset{\text{def}}{=} \lambda_0 \leq 2 \sup_{a \in \Delta} \|a\|.$$

If we discard the first n terms of our series, then λ_0 can decrease, but it does not tend to zero as $n \to \infty$ because the series is divergent. By taking this into account (and by discarding if necessary finitely many terms of the series) we find a sequence of non-overlapping partial sums a_n for which $\lambda(a_n) \geq 3\lambda_0/4$. We set $g_n = \sum \alpha_i^n e_i^n$, where the e_i^n are the elements in the sum a_n, and the α_i^n are chosen so that

$$\|g_n\| = \lambda(a_n).$$

We introduce a new series $\sum g_n$. From our construction we immediately obtain the following important properties:

$$\|\sum \xi_i g_i\| \leqslant \lambda_0 \max\{|\xi_i|\}, \tag{6}$$

$$\|g_i\| \geqslant 3\lambda_0/4, \tag{7}$$

where $\{\xi_i\}$ is a sequence of real numbers that are non-zero only for finitely many indices i.

For any functional $y \in X^*$ the series $\sum y(g_n)$ is absolutely convergent. From the estimate (6) we obtain

$$\sum_{n=1}^{\infty} |y(g_n)| \leqslant \lambda_0 \|y\|. \tag{8}$$

We consider the element $\kappa = \sum \xi_i g_i$. Let $\xi_k = \pm\max\{|\xi_i|\}$. We choose a functional $y \in X^*$ for which $y(g_k) = \mp 1$, $\|y\| = 1/\|g_k\|$. Then we have

$$\|y\|\,\|\kappa\| \geqslant y(\kappa) = |\alpha_k| - \sum_{i \neq k} |y(g_i)|\,|\alpha_i|$$

$$\geqslant \max\{|\alpha_i|\}\left(1 - \sum_{i \neq k} |y(g_i)|\right).$$

From the inequalities (7) and (6) we obtain the estimate

$$\sum_{i \neq k} |y(g_i)| \leqslant \lambda_0 \|y\| - 1 \leqslant \tfrac{4}{3} - 1 = \tfrac{1}{3}.$$

Hence, finally,

$$\lambda_0 \|\kappa\| \geqslant \tfrac{2}{3} \max\{|\alpha_i|\}. \tag{9}$$

Now we consider the linear manifold of all finite sums of the form $\kappa = \sum \alpha_i g_i$. The estimates (6) and (9) show that on this manifold the original norm and the norm $\|\kappa\|_1 = \max\{|\alpha_i|\}$ are equivalent, and hence the closure of this manifold is isomorphic to c_0, as we required.

We make another observation about the divergent unconditionally bounded series $\sum_{n=1}^{\infty} e_n$. We prove that we can find a series of the form $\sum_{i=1}^{\infty} e_{n_i}$ that is not weakly convergent to an element in X. Assume that this is not so. Then it is easy to see that the set Δ must be weakly compact. We choose a finite sum $\kappa = \sum \alpha_i e_i$, where $|\alpha_i| \leqslant 1$. We rearrange the terms in this sum so that the coefficients α_i are non-increasing. Then by using Abel's identity we see that κ belongs to the 'doubled' convex hull of the set Δ.

We now consider the series $\sum_{n=1}^{\infty} g_n$ introduced earlier. From its construction we obtain

$$\sum_{n=1}^{\infty} g_n = \sum_{i=1}^{\infty} \alpha_i e_i,$$

where $|\alpha_i| \leq 1$. Hence the series must converge weakly to an element in X. But this is impossible, since under the isomorphism, g_n is mapped into the nth unit vector in c_0.

2. Let $f(t)$ be an almost periodic function and $u(t) = \int_0^t f(s)\,ds$. It is clear that for each $a \in J$ the difference $u(t+a) - u(t) = \int_t^{t+a} f(s)\,ds$ is an almost periodic function. Therefore the problem concerning the almost periodicity of the integral is a particular case of the following problem on differences.

Let G be an arbitrary group with the group operation written multiplicatively. Let $u: G \to X$ be a function on the group. We assume that for each $\gamma \in G$ the difference

$$u(t\gamma) - u(t) = g_\gamma(t) \tag{10}$$

is an almost periodic function of $t \in G$, and pose the question, is $u(t)$ almost periodic? The problem can be extended in a sense by taking the function $g_\gamma(t)$ in a wider class, say the class of N-almost periodic functions or the class of recurrent functions on the group G, and asking whether $u(t)$ belongs to the appropriate class.

We solve the problem on differences under the assumption that one of the following two conditions holds:

(a) the function $u(t)$ is bounded and the space X does not contain c_0;

(b) $u(t)$ is weakly compact.

Let e be the unit element of G. We set, by definition,

$$U(\varepsilon) = \{\tau : \|u(\tau) - u(e)\| < \varepsilon\},$$
$$V_\gamma(\delta) = \{\tau : \|g_\gamma(\tau) - g_\gamma(e)\| < \delta\}.$$

Lemma 1. *For any $\varepsilon > 0$ there is a $\delta > 0$ and a finite collection of elements $\gamma_1, \ldots, \gamma_r$ such that*

$$U(\varepsilon) \supset \bigcap_{i=1}^{r} V_{\gamma_i}(\delta). \tag{11}$$

Proof. Relation (10) determines $u(t)$ up to a constant; therefore we can assume that $u(e) = 0$.

Suppose that (11) does not hold. Then

$$\bigcap_{i=1}^{r} V_{\gamma_i}(\delta) \cap G \backslash U(\varepsilon_0) \neq \varnothing, \tag{12}$$

for some $\varepsilon_0 > 0$, any $\delta > 0$ and any collection $\gamma_1, \ldots, \gamma_r \in G$.

We show that we can form a sequence of elements $t_n \in G$ for which

$$\|u(t_n)\| \geqslant \varepsilon_0 \quad (n \geqslant 1), \tag{13}$$

$$\|u(t_n t_{i_k} \cdots t_{i_1}) - u(t_n) - u(t_{i_k} \cdots t_{i_1})\| \leqslant \varepsilon_0/2^n \quad (n \geqslant 2), \tag{14}$$

where the $i_1, \ldots, i_k$ are integers such that

$$1 \leqslant i_1 < i_2 < i_3 < \cdots < i_k \leqslant n-1. \tag{15}$$

We construct the sequence $\{t_n\}$ by induction. It follows from our assumption that there exists an element t_1 such that $t_1 \neq U(\varepsilon_0)$, that is, $\|u(t_1)\| \geqslant \varepsilon_0$. Suppose that $t_1, \ldots, t_{n-1}$ have been constructed. We consider the set

$$K_n(\varepsilon_0) = \bigcap V_{t_{i_k} \cdots t_{i_1}}(\varepsilon_0/2^n),$$

where the intersection is taken over all collections of integers $i_1, \ldots, i_k$ satisfying (15). It follows from (12) that we can find an element $t_n \in K_n(\varepsilon_0)$ such that $t_n \notin U(\varepsilon_0)$. Since

$$g_\gamma(t_n) - g_\gamma(e) = u(t_n\gamma) - u(t_n) - u(\gamma),$$

the required inequality (14) is equivalent to the condition $t_n \in K_n(\varepsilon_0)$. This establishes the possibility of constructing the sequence $\{t_n\}$.

We consider the series $\sum_{n=1}^{\infty} u(t_n)$. Let $\tau_k = t_{n_k} (k = 1, \ldots, m)$ be any terms of it such that $n_1 < n_2 < \cdots < n_m$. Then we have the identity

$$\begin{aligned} u(\tau_m \cdots \tau_1) - \sum_{i=1}^{m} u(\tau_i) = {} & [u(\tau_m \cdots \tau_1) - u(\tau_m) - u(\tau_{m-1} \cdots \tau_1)] \\ & + [u(\tau_{m-1} \cdots \tau_1) - u(\tau_{m-1}) - u(\tau_{m-2} \cdots \tau_1)] \\ & + \cdots + [u(\tau_2\tau_1) - u(\tau_2) - u(\tau_1)], \end{aligned}$$

from which we find by (14) that

$$\left\| \sum_{i=1}^{m} u(\tau_i) \right\| = \|\sum u(t_{n_i})\| \leqslant \sup_{t \in G} \|f\| + \varepsilon_0.$$

The last result together with (13) means that $\sum_{n=1}^{\infty} u(t_n)$ is a divergent unconditionally bounded series, and so our lemma is proved in the case of condition (a).

Now suppose that condition (b) holds. We are going to prove that every series of the form $\sum u(t_{n_k}) = \sum u(\tau_k)$ converges weakly to an element in X. As we observed in § 2.1 (subsection 1 of § 2) this is impossible for divergent series.

Let

$$s_k = \sum_{i=1}^{k} u(\tau_i) \quad \text{and} \quad \tilde{s}_k = u(\tau_k \cdots \tau_1).$$

We prove that the sequence $\{s_k - \tilde{s}_k\}$ is fundamental. In fact, since for $m > n$ we have

$$s_n - \tilde{s}_n - s_m + \tilde{s}_m = u(\tau_m \cdots \tau_1) - \sum_{i=n}^{m} u(\tau_i) - u(\tau_n \cdots \tau_1)$$
$$= [u(\tau_m \cdots \tau_1) - u(\tau_m) - u(\tau_{m-1} \cdots \tau_1)]$$
$$+ [u(\tau_{m-1} \cdots \tau_1) - u(\tau_{m-1}) - u(\tau_{m-2} \cdots \tau_1)]$$
$$+ \cdots + [u(\tau_{n+1} \cdots \tau_1) - u(\tau_{n+1}) - u(\tau_n \cdots \tau_1)],$$

then from the inequalities (14) we obtain

$$\|s_n - \tilde{s}_n - s_m + \tilde{s}_m\| < \varepsilon_0/2^{n-1}.$$

Because the sequence $\tilde{s}_k$ is weakly compact (see condition (b)), the sequence s_k is also weakly compact. Hence the lemma is completely proved.

The lemma has several corollaries.

Corollary 1. *If for any collection $\gamma_1, \ldots, \gamma_r \in G$ the set $\bigcap V_{\gamma_i}(\delta)$ is relatively dense in G, then $u(t)$ is compact.*

Proof. Let D be the set of values of $u(t)$. By the lemma, the set $U(\varepsilon)$ is relatively dense; therefore we can find elements $s_1, \ldots, s_n \in G$ such that $G = \bigcup U(\varepsilon)\, s_i$. Then $D = \bigcup D_i$, where $D_i = \{u(\tau s_i) \colon \tau \in U(\varepsilon)\}$. We can assume that $u(e) = 0$. We consider the identity

$$u(ts_i) - g_{s_i}(t) = u(t).$$

Since $\|u(t)\| \leqslant \varepsilon$ for $t \in U(\varepsilon)$, the set $\{g_{s_i}(t)\}$ is an ε-net for the set D_i. Because the functions $g_{s_i}(t)$ are compact, $\bar{D}$ is compact.

For simplicity, the other corollaries are stated for the case $G = J$.

Theorem 2. *Suppose that one of conditions (a) or (b) holds. Then the indefinite integral of an almost periodic (respectively, recurrent) function is almost periodic (respectively, recurrent).*[1]

Proof. It is sufficient to establish that $u(t)$ is compact. It is clear that the function $g_a(t) = \int_t^{t+a} f(s)\, ds$ is recurrent for any $a \in J$. Furthermore, for any finite collection $a_1, \ldots, a_r \in J$ the set $\bigcap_{i=1}^r V_{a_i}(\delta)$ is relatively dense since it contains the set $L(\delta/2N, N, f)$, where $N = \max |a_i|$. It is now enough to refer to Corollary 1.

Corollary 2. *Suppose that a function $f(t): J \to X$ is Poisson stable, that is, it has a 'returning' sequence $\{t_m\}$. Then $\{t_m\}$ is also 'returning' for $u(t)$.*

[1] Another proof of Theorem 2 in the case when condition (b) holds is given in Chapter 8, § 5.

Proof. From the equality (2) it is clear that it is enough to prove that $\|u(t_m)\| \to 0$. The convergence $f(t+t_m) \xrightarrow{\text{loc}} f(t)$ means that $t_m \in V_s(\delta)$ for $m \geq m(s, \delta)$. Then from Lemma 1 we obtain that $t_m \in U(\varepsilon)$ as we required.

3. Information from harmonic analysis

Let $\mathscr{L}$ be a collection of measurable complex valued functions $\phi(t)$ $(t \in J)$ with the norm $\|\phi\|_{\mathscr{L}} = \int_J |\phi(t)|\, dt$, and let $\hat{\mathscr{L}}$ denote the collection of Fourier transforms $\hat{\phi}(\lambda) = \int_J \phi(t) \exp(-i\lambda t)\, dt$. We denote by C_0^∞ the subset of $\hat{\mathscr{L}}$ consisting of infinitely differentiable functions with compact support, and the support of $\hat{\phi}(\lambda) \in C_0^\infty$ by $\operatorname{supp} \hat{\phi}$.

We fix a $\hat{\phi}_0(\lambda) \in C_0^\infty$ such that $\hat{\phi}_0(\lambda) = 1$ for $|\lambda| \leq 1$ and $\operatorname{supp} \hat{\phi}_0 \subset [-2, 2]$. For every $a > 0$ we set

$$\hat{K}_a(\lambda) = \hat{\phi}_0(\lambda/a) \quad \text{and} \quad K_a(t) = a\phi_0(at).$$

We recall that $C = C(X)$ stands for the set of all bounded continuous functions $J \to X$, and $\mathring{C} = \mathring{C}(X)$ for the subspace of $C(X)$ consisting of almost periodic functions. In the space $C(X)$, besides norm convergence, a local convergence can be defined: $f_n \xrightarrow{\text{loc}} f$ if the sequence $\{f_n\}$ is bounded in $C(X)$ and $f_n(t) \to f(t)$ uniformly on each finite interval. In particular, the set of trigonometric polynomials is locally dense in $C(X)$.

1. A point $\lambda \in J$ is called ***regular*** for the function $f \in C(X)$ if we can find a neighbourhood of it such that

$$f * \phi = \int_J f(t-s)\phi(s)\, ds \equiv 0$$

for every $\hat{\phi} \in C_0^\infty$ with support in this neighbourhood. The set of all regular points is obviously open; its complement is called the *spectrum* of f and is denoted by $\sigma = \sigma(f)$.

We can deduce directly from the definition a number of important properties of the spectrum. First of all, the spectrum is not empty provided that $f \not\equiv 0$. For if we assume that $\sigma(t) = \varnothing$, then we can decompose any element $\hat{\phi} \in C_0^\infty$ into a sum of a finite number of elements with sufficiently small supports. Then we obtain that $f * \phi \equiv 0$. But since the following proposition holds, we then have $f \equiv 0$.

Proposition 3. *For $f \in C(X)$ we set $f_n = f * K_n$. Then $f_n \xrightarrow{\text{loc}} f$ as $n \to \infty$.*

If f is uniformly continuous, then $f_n \xrightarrow{J} f$.

Proof. We consider the case of uniform continuity; the changes necessary for the general case will be self-evident.

Since $\int_J K_n(t)\,dt = \hat{K}_n(0) = 1$, we have

$$\begin{aligned} f(t) - f_n(t) &= \int_J [f(t) - f(s)]K_n(t-s)\,ds \\ &= -\int_J [f(t) - f(t-u)]K_n(u)\,du \\ &= -\int_J [f(t) - f(t-u/n)]\phi_0(u)\,du. \end{aligned}$$

We choose a number $u_0 = u_0(\varepsilon)$ for which $\int_{|u|\geq u_0} |\phi_0(u)|du \leq \varepsilon$. Then by uniform continuity we have

$$\|f(t) - f(t-u/n)\| \leq \varepsilon \quad (n \geq N(\varepsilon), t \in J).$$

But then for any $t \in J$ we have

$$\|f(t) - f_n(t)\| \leq \varepsilon\|\phi_0\|_{\mathscr{L}} + 2\varepsilon\|f\|_C,$$

as we required.

Now we consider a function $f \in C(X)$ with a compact spectrum. Suppose that $\hat{\phi}(\lambda) \in C_0^\infty$ is such that $\hat{\phi}(\lambda) = 1$ for $\lambda \in \sigma(f)$. It is not difficult to see that the spectrum of $f - f * \phi$ is empty, that is,

$$f(t) = f * \phi = \int_J f(s)\phi(t-s)\,ds. \tag{16}$$

The function ϕ (being the inverse Fourier transform of the function $\hat{\phi}$ of compact support) belongs to the class $\mathscr{L}$ together with all its derivatives. Therefore, f belongs to $C(X)$ together with all its derivatives.

We note another completely obvious fact. Suppose that the spectrum of $f \in C(X)$ lies outside an interval $[a, b]$. Then we can find a sequence of trigonometric polynomials f_n such that

$$f_n \xrightarrow{\text{loc}} f, \quad \sigma(f_n) \subset J \backslash [a, b].$$

To prove this we must first choose an arbitrary sequence of polynimials g_n for which $g_n \xrightarrow{\text{loc}} f$, and then 'cut off' the part of the spectrum lying in the interval $[a, b]$. This means the following. The set $\sigma(f)$ obviously lies outside some interval $[a-\varepsilon, b+\varepsilon]$. Therefore we can set $f_n = g_n - g_n * \psi$, where the function ψ is such that $\hat{\psi}(\lambda) = 1$ for $\lambda \in [a, b]$ and $\hat{\psi}(\lambda) = 0$ for $\lambda \notin [a-\varepsilon, b+\varepsilon]$.

2. For what follows we shall need the following lemma.

Lemma 2. *Suppose that a function $f \in C(X)$ is such that the mean*

$$\lim_{T\to\infty} \frac{1}{2T}\int_{-T+\alpha}^{T+\alpha} f(t)\,\mathrm{d}t = 0$$

exists and is attained uniformly with respect to $\alpha \in J$. We set $f_\varepsilon = f * K_\varepsilon$ and $f^\varepsilon = f - f_\varepsilon$. Then $\|f - f^\varepsilon\|_C \to 0$ as $\varepsilon \to 0$.

Proof. We integrate by parts the expression $f_\varepsilon = \int_J f(t-s)\varepsilon\phi_0(\varepsilon s)\,\mathrm{d}s$:

$$\begin{aligned} f_\varepsilon(t) &= \varepsilon^2 \int_J \int_0^s f(t-u)\,\mathrm{d}u\,\phi_0'(\varepsilon s)\,\mathrm{d}s \\ &= \int_J \frac{\varepsilon}{\eta}\int_0^{\eta/\varepsilon} f(t-u)\,\mathrm{d}u\,\eta\phi_0'(\eta)\,\mathrm{d}\eta \\ &= \int_{|\eta|\leq\sqrt{\varepsilon}} \frac{\varepsilon}{\eta}\int_0^{\eta/\varepsilon} f(t-u)\,\mathrm{d}u\,\eta\phi_0'(\eta)\,\mathrm{d}\eta \\ &\quad + \int_{|\eta|\geq\sqrt{\varepsilon}} \frac{\varepsilon}{\eta}\int_0^{\eta/\varepsilon} f(t-u)\,\mathrm{d}u\,\eta\phi_0'(\eta)\,\mathrm{d}\eta \\ &= \bar{f}_\varepsilon + \bar{\bar{f}}_\varepsilon. \end{aligned}$$

The term $\bar{f}_\varepsilon$ is estimated easily since

$$\|\bar{f}_\varepsilon\|_C \leq \sqrt{\varepsilon}\|f\|_C \sup_{\eta\in J} |\eta\phi_0'(\eta)|.$$

For the estimation of $\bar{\bar{f}}_\varepsilon$ we observe that from the uniformness of the mean it follows that the convergence

$$\frac{\varepsilon}{\eta}\int_0^{\eta/\varepsilon} f(t-u)\,\mathrm{d}u \to 0$$

is uniform with respect to $t \in J$, $|\eta| \geq \sqrt{\varepsilon}$. Therefore, $\|\bar{\bar{f}}_\varepsilon\|_C \to 0$ as we required to prove.

The function f^ε in Lemma 2 has the property that its spectrum is separated from 0. The following important theorem holds for functions whose spectrum is separated from 0.

Theorem 3. *If the spectrum of $f \in C(X)$ is separated from 0, then the indefinite integral $\int_0^t f(s)\,\mathrm{d}s$ is an element of $C(X)$. Here if $f \in \mathring{C}(X)$, then $\int_0^t f(s)\,\mathrm{d}s \in \mathring{C}(X)$.*

Proof. Suppose that the set $\sigma(f)$ lies outside the interval $[-\alpha, \alpha]$. We choose a twice continuously differentiable function $\hat{\phi}(\lambda)$ such that $\hat{\phi}(\lambda) = 1/\mathrm{i}\lambda$ for $|\lambda| \geq \alpha$. We are going to show that the inverse transform $\phi(t)$ belongs to the class $\mathscr{L}$.

We integrate $\phi(t)=(1/2\pi)\int_J \hat{\phi}(\lambda)\exp(i\lambda t)\,d\lambda$ twice by parts; then we obtain

$$|\phi(t)| \leqslant \frac{1}{2\pi}\int_J \left|\frac{\hat{\phi}''(\lambda)\exp(i\lambda t)}{it^2}\right| d\lambda$$
$$=\frac{1}{2\pi t^2}\int_J |\hat{\phi}''(\lambda)|\,d\lambda. \tag{17}$$

In addition we have

$$\phi(t)=\frac{1}{2\pi}\int_{-\alpha}^{\alpha}\hat{\phi}(\lambda)\exp(i\lambda t)\,d\lambda+\frac{1}{\pi}\int_{\alpha}^{\infty}\frac{\sin\lambda t}{\lambda}\,d\lambda.$$

Therefore,

$$\phi(t)=\frac{1}{2\pi}\int_{-\alpha}^{\alpha}\hat{\phi}(\lambda)\exp(i\lambda t)\,d\lambda + \frac{1}{\pi}\int_{\alpha t}^{\infty}\frac{\sin\eta}{\eta}\,d\eta \quad (t>0)$$
$$\phi(t)=\frac{1}{2\pi}\int_{-\alpha}^{\alpha}\hat{\phi}(\lambda)\exp(i\lambda t)\,d\lambda - \frac{1}{\pi}\int_{-\alpha t}^{\infty}\frac{\sin\eta}{\eta}\,d\eta \quad (t<0).$$

Hence it is clear that $\phi(t)$ is continuous everywhere apart from $t=0$, where it has a discontinuity of the first kind ($\phi(+0)-\phi(-0)=1$). Consequently, by (17) $\phi\in\mathcal{L}$.

We set

$$u(t)=\int_J f(t)\phi(t-s)\,ds$$
$$=\int_J f(t-s)\phi(s)\,ds.$$

Since it is clear that $u\in C(X)$, it is sufficient to prove that $u'=f$. With this aim we assume that f is a trigonometric polynomial $f=\sum a_m\exp(i\lambda_m t)$. Since $|\lambda_m|\geqslant\alpha$ we have

$$u(t)=\sum a_m\exp(i\lambda_m t)\int_J \exp(-i\lambda_m s)\phi(s)\,ds$$
$$=\sum a_m\exp(i\lambda_m t)\hat{\phi}(\lambda_m)=\sum\frac{a_m}{i\lambda_m}\exp(i\lambda_m t),$$

that is, $u'=f$.

Now we choose a sequence of polynomials $f_n(t)$ for which

$$f_n \xrightarrow{\text{loc}} f, \quad \sigma(f_n) \subset J \backslash [-\alpha, \alpha].$$

If we set $u_n = f_n * \phi$, then $u_n \xrightarrow{\text{loc}} u$ and $u_n' = f_n$. Hence it follows that $u' = f$, and the theorem is proved.

4. A spectral condition for almost periodicity

Suppose that we know beforehand that $f(t)$ belongs to the class $C(X)$. It is often possible to establish the almost periodicity of this function starting only from certain *a priori* information about its spectrum. For example, if the spectrum of $f(t)$ is a finite set, then $f(t)$ is a trigonometric polynomial. To prove this fact note that $f(t)$ has a representation (16) in which ϕ is such that its Fourier transform $\hat{\phi}$ is concentrated only in a small neighbourhood of the spectrum. Let $\sigma(f) = \{\lambda_i\}$. We can partition $\hat{\phi}$ into a sum of a finite number of functions $\hat{\phi}_i$ which are non-zero only in a small neighbourhood of the point λ_i. Then the problem is reduced to a study of a function with a spectrum at a single point. By multiplying by $\exp(-i\lambda_i t)$ the spectrum can be moved to 0. Thus, we will prove that if $\sigma(f) = \{0\}$, then $f \equiv \text{const}$. For this we consider the equality $f' = f * K_\varepsilon'$. Since

$$\int_J |K_\varepsilon'(t)| \, dt = \varepsilon^2 \int_J |\phi_0'(\varepsilon t)| \, dt$$

$$= \varepsilon \int_J |\phi_0'(t)| \, dt \to 0,$$

we have

$$\|f'\|_C \leq \|f\|_C \|K_\varepsilon'\|_{\mathscr{L}} \to 0,$$

that is, $f'(t) \equiv 0$.

This simple result about functions with a finite spectrum has a fairly significant generalisation, but before discussing it we introduce a new concept.

A point $\lambda \in J$ is called a *point of almost periodicity* of a function $f \in C(X)$ if it has a neighbourhood such that for any $\hat{\phi} \in C_0^\infty$ with support in this neighbourhood the convolution $f * \phi$ is an almost periodic function. The set of all points of almost periodicity is, obviously, open. Its complement is called the *set of points of non-almost periodicity of* f and is denoted by $\Delta = \Delta(f)$.

Clearly, $\Delta(f) \subset \sigma(f)$. In addition, the following property holds: if $f \in C(X)$ is uniformly continuous and $\Delta(f)$ is empty, then f is almost periodic. In fact, from $\Delta(f)=\varnothing$ it easily follows that $f*\phi \in \mathring{C}(X)$ for any $\hat{\phi} \in C_0^\infty$, in particular, $f_n = f*K_n \in \mathring{C}(X)$. It remains to refer to Proposition 3.

We say that a closed subset of the real line is *rarified* if it does not contain non-empty perfect subsets (for instance, if it is countable).

Theorem 4. *Let $f \in C(X)$ be uniformly continuous and $\Delta(f)$ be rarified (for example, its spectrum is rarified). Then $f \in \mathring{C}(X)$ if one of the following conditions holds:*

(a) the space X does not contain c_0;

(b) f is weakly compact.

Proof. Let us assume that $\Delta = \Delta(f) \neq \varnothing$. Since Δ is rarified it has an isolated point λ_0. By the definition of the points of non-almost periodicity, we can find a $\hat{\phi} \in C_0^\infty$ such that

$$\operatorname{supp} \hat{\phi} \cap \Delta = \{\lambda_0\}, \quad f_1 = f*\phi \notin \mathring{C}(X).$$

By multiplying f by $\exp(-\lambda_0 t)$ we can make $\lambda_0 = 0$. It is easily seen that $\Delta(f_1)=\{0\}$. Now f_1 is differentiable because it has a compact spectrum. We set $g(t)=f_1'(t)$ and will prove that $g \in \mathring{C}(X)$. Since g is the derivative of a bounded function its uniform mean is zero. Therefore, by Lemma 2 $g_\varepsilon = g*K_\varepsilon \xrightarrow{J} 0$. On the other hand, the set $\Delta(g^\varepsilon)$ for $g^\varepsilon = g - g_\varepsilon$ is empty. In fact, $\Delta(g^\varepsilon)$ can consist at most of 0, but the spectrum of g^ε is separated from 0. Hence $g^\varepsilon \in \mathring{C}(X)$, but then $g \in \mathring{C}(X)$.

Thus, $f_1'(t) \in \mathring{C}(X)$. Then by the theorem on the integral (Theorem 2)[2] we obtain that $f_1(t) \in \mathring{C}(X)$, which contradicts our initial assumption.

5. Harmonic analysis of bounded solutions of linear equations

The spectral condition for almost periodicity obtained in § 4 (Theorem 4) allows us to investigate the structure of bounded solutions of the linear equation $u' = Au = f$ with a constant operator A.

Suppose that an operator A (in general unbounded) is the generating operator of a strongly continuous semigroup $U(t)$ $(t \geqslant 0)$. We

[2] In the case of condition (*b*) it should be noted that the functions f_1 and f_1' are weakly compact along with f since they are expressed in terms of f as convolutions.

consider the non-homogeneous equation

$$Lu = u' - Au = f, \tag{18}$$

where the free term f is almost periodic.

We are interested in solutions that are defined and bounded on the whole real line. A function $u : X \to J$ is defined to be a solution if

$$u(t) = U(t-t_0)u(t_0) + \int_{t_0}^{t} U(t-s)f(s)\,\mathrm{d}s$$

holds for every $t \geq t_0 \in J$. It is easily seen that $u(t)$ is continuous on the whole line.

We denote by $\sigma_0(A)$ the intersection of the spectrum of the operator A with the imaginary axis, by $\Delta(u)$ the set of points of non-almost periodicity of a bounded solution $u(t)$, and by Hom (X, X) the set of all bounded linear operators $X \to X$ with the operator norm.

Lemma 3. $\Delta(u) \subset \sigma_0(A)$.

Proof. Let $\mathrm{i}\lambda_0$ $(\lambda_0 \in J)$ be a regular point for A; the resolvent $R_\lambda = (A - \mathrm{i}\lambda)^{-1}$ is analytic in some neighbourhood $(-4\alpha + \lambda_0, 4\alpha + \lambda_0)$.

We fix a function $\hat{\phi}(\lambda) \in C_0^\infty$ such that $\hat{\phi}(\lambda) = 1$ for $\lambda \in [-2\alpha + \lambda_0, 2\alpha + \lambda_0]$, and supp $\hat{\phi} \subset (-4\alpha + \lambda_0, 4\alpha + \lambda_0)$. Since $\hat{\phi}(\lambda)R_\lambda : J \to$ Hom (X, X) is infinitely differentiable and has compact support, it is the Fourier transform of a continuous function $F(t) : J \to$ Hom (X, X) and $\int_J \|F(t)\|\,\mathrm{d}t < \infty$, that is, $\hat{\phi}(\lambda)R_\lambda = \int F(t)\exp(-\mathrm{i}\lambda t)\,\mathrm{d}t$.

Let $\hat{\psi}$ be any element in C^∞ for which supp $\hat{\psi} \subset [-\alpha + \lambda_0, \alpha + \lambda_0]$. By setting $v = u * \psi$ and $g = f * \psi$ we obtain $Lv = v' + Av = g$. The function $v(t)$ has a compact spectrum, and therefore $v'(t) \in C(X)$. It is easy to see from here that the equation $v' + Av = g$ is satisfied in the ordinary sense, that is, $Av \in C(X)$.

Now we choose a sequence of trigonometric polynomials v_n such that

$$v_n \xrightarrow{\text{loc}} v, \quad Av_n \xrightarrow{\text{loc}} Av, \quad v'_n \xrightarrow{\text{loc}} v'.$$

Then we have $Lv_n = g_n$, where $g_n \xrightarrow{\text{loc}} g$. Since the spectrum of the function $v(t)$ is contained in the interval $[-\alpha + \lambda_0, \alpha + \lambda_0]$ we can suppose that the spectrum of the polynomial v_n is contained in $[-2\alpha + \lambda_0, 2\alpha + \lambda_0]$.

We prove that

$$v_n(t) = \int_J F(s)g_n(t-s)\,\mathrm{d}s \tag{19}$$

Let $g_n(t)=\sum_k a_k \exp(\mathrm{i}\lambda_k t)$. Since $\lambda_k\in[-2\alpha+\lambda_0, 2\alpha+\lambda_0]$ and so $\hat{\phi}(\lambda_k)=1$, we have

$$\begin{aligned}\int_J F(s)g_n(t-s)\,\mathrm{d}s &= \sum_k \exp(\mathrm{i}\lambda_k t)\int_J F(s)\exp(-\mathrm{i}\lambda_k s)a_k\lambda\,\mathrm{d}s\\ &= \sum \phi(\lambda_k)R_{\lambda_k}a_k \exp(\mathrm{i}\lambda_k t)\\ &= \sum (A-\mathrm{i}\lambda_k)^{-1}a_k \exp(\mathrm{i}\lambda_k t)=v_n(t)\end{aligned}$$

which proves (19). By taking the limit in (19) we obtain

$$\begin{aligned}v(t) &= \int_J u(s)\psi(t-s)\,\mathrm{d}s\\ &= \int_J F(s)g(t-s)\,\mathrm{d}s,\end{aligned}$$

from which it follows that $v(t)\in\mathring{C}(X)$. But the latter means that $\lambda_0\notin\Delta(u)$, and so the lemma is proved.

The next result is obtained from this lemma and Theorem 4.

Theorem 5. *Suppose that the intersection of the spectrum of the operator A and the imaginary axis is a rarified set. Then every bounded solution is weakly almost periodic. If, in addition, one of conditions (a) and (b) of Theorem 4 holds, then every bounded uniformly continuous solution is almost periodic.*

Proof. The second part of the theorem is obtained directly from Lemma 3 and Theorem 4. To prove the first part we need to verify the weak uniform continuity of a bounded solution.

Let $u(t)$ be a bounded solution. Clearly, it is enough to verify the uniform continuity of the function $(y, u(t))$, where y ranges over a set dense in X^*. But if we choose y from the domain of the adjoint operator A^*, then we obtain

$$(y, u')=-(u, A^*y)+(y, f),$$

from which it follows that $(y, u(t))$ is uniformly continuous.

Let us note that in the general case (when the operator A is not bounded), a bounded solution is not necessarily uniformly continuous.

Example. We consider the equation

$$x^{(m)}+\sum_{k=0}^{m-1} A_k(t)x^{(k)}=f(t), \tag{20}$$

Here $f \in \mathring{C}(X)$, and the $A_k(t)$ $(k = 0, \ldots, m-1)$ are continuous periodic (with a common period) functions $J \to \text{Hom}(X, X)$. We assume that the operator $A_k(t)$ $(t \in J, k = 0, \ldots, m-1)$ is compact and that the space X does not contain c_0. We are going to prove that every bounded solution $x(t)$ is almost periodic together with its derivatives of orders up to m.

We set $u(t) = \{x(t), x'(t), \ldots, x^{(m-1)}(t)\}$ and write (26) in the form of a first order equation: $u' + A(t)u = g$. As will be shown below (see Proposition 4), the vector-function $u(t)$ is bounded. The operator $A(t)$ has the property that its mth power is compact. Hence our equation reduces to one of the form $v' + Qu = \tilde{g}$, where Q does not depend on t and Q^m is a compact operator. Since the spectrum of Q is rarefied, the matter is reduced to Theorem 5.

Proposition 4. *If $x(t)$ is a bounded solution of the equation, then the derivatives $x^{(k)}$ $(k = 1, \ldots, m)$ are also bounded.*

Proof. We begin by deducing a certain inequality, first for real and then for abstract functions.

Let $g(t)$ be a real twice continuously differentiable function on $[-T, T]$. Then

$$\sup_{|t| \leq T} |g'(t)|^2 \leq 4 \sup_{|t| \leq T} |g(t)| \sup_{|t| \leq T} |g''(t)| + \inf_{|t| \leq T} |g'(t)|^2. \tag{21}$$

To prove this we choose a point $t_1 \in [-T, T]$ for which $|g'(t_1)|^2 = \sup_{|t| \leq T} |g'(t)|^2$. We take t_0 to be the nearest point to t_1 for which $|g'(t_0)|^2 = \inf_{|t| \leq T} |g'(t)|^2$. It is clear that $g'(t)$ does not change sign on $[t_0, t_1]$. Since $\mathrm{d}(g')^2/\mathrm{d}t = 2g''g'$, by the generalised mean value theorem we have

$$\begin{aligned} |g'(t_1)|^2 - |g(t_0)|^2 &= 2\int_{t_0}^{t_1} g''g' \, \mathrm{d}t \\ &= 2g''(\xi)\int_{t_0}^{t} g' \, \mathrm{d}t \\ &= 2g''(\xi)(g(t_1) - g(t_0)), \end{aligned}$$

after which the inequality (21) is obvious.

Now let $x(t)$ be an abstract function $[-T, T] \to X$, where X is a real space, and let $y \in X^*$ be a real functional with unit norm. We apply (21) to $g(t) = (y, x(t))$. Then we obtain for any $t \in [-T, T]$

$$|(y, x'(t))|^2 \leqslant 4 \sup_{|t| \leqslant T} |(y, x(t)) \sup_{|t| \leqslant T} |(y, x''(t))| + \inf_{|t| \leqslant T} (y, x'(t))^2.$$

We choose $y \in X^*$ so that $(y, x'(t)) = \|x'(t)\|$. Then

$$\sup_{|t| \leqslant T} \|x'(t)\|^2 \leqslant 4 \sup_{|t| \leqslant T} \|x(t)\| \sup_{|t| \leqslant T} \|x''(t)\| + \inf_{|t| \leqslant T} \|x'(t)\|^2. \tag{22}$$

By considering X as a space with real scalars, (22) can be established in the general case.

After these remarks we return to Proposition 4. Suppose that the functions $x(t), x^{(1)}(t), \ldots, x^{(k)}(t)$ $(k \geqslant 0)$ are bounded but $x^{(k+1)}(t)$ is unbounded. From the inequality (22) it follows that $x^{(k+2)}(t)$ must grow faster than $x^{(k+1)}$ in the sense that

$$(\sup_{|t| \leqslant T} \|x^{(k+1)}(t)\|)/(\sup_{|t| \leqslant T} \|x^{(k+2)}(t)\|) = o(1) \quad \text{as } T \to \infty.$$

Hence, in turn, we find that $x^{(k+3)}(t)$ grows faster than $x^{(k+2)}(t)$ and so on. By repeating this process, we obtain that $x^{(m)}(t)$ grows faster than any of the lower order derivatives. But by (20), $x^{(m)}(t)$ is majorised by a linear combination of the lower order derivatives.

Comments and references to the literature

§ **1.** The theorem concerning the integral was proved by Bohl [16] for the case of scalar conditionally periodic functions. Bohr established that Bohl's proof is suitable for the case of almost periodic functions. The important generalisation to the case of functions with values in a uniformly convex space is due to Amerio (see the book by Amerio & Prouse [2]). Functions stable in the sense of Poisson have been studied systematically by Shcherbakov [115]. The proof of Proposition 1 is based on an elegant idea of Bochner [28].

§ **2.** Banach spaces that do not contain c_0 were introduced by Pełczynski [96]; in particular, Proposition 2 is due to him. The important role of these spaces in the problem of integration was discovered by Kadets [63] and [64], who proved Theorem 2 for almost periodic functions on the line. The problem about differences on a group and also the generalisation of the theorem concerning the integral to recurrent and Poisson stable functions is due to Boles [11]. Certain results on integration (close to the individual results of Boles) were obtained by Shcherbakov [116] and Bronshtein (unpublished). We mention that the work of Boles was preceded by the article of Zhikov [51], who considered the problem of integrating recurrent and N-almost periodic functions. On the whole, our exposition in § 2 follows the treatment of Boles.

§ 3. Lemma 2 (in a rather veiled form) appears in an article by Levitan [77]. It can be shown that Lemma 2 is equivalent to the well-known theorems of Bogolyubov on approximations, which play an important role in justifying the averaging principle [8]. In Chapter 11 we apply Lemma 2 in justifying the averaging principle. Theorem 3 is due to Favard [106].

§4. The spectral condition for almost periodicity for scalar functions was stated by Loomis [82] whose work was preceded by the articles of Levitan [75] and Wolf [35]. The generalisation to abstract functions is not trivial. For the case of a Hilbert space, Zhikov [47] has given certain conditions for almost periodicity. Further generalisations are due to Boles [12] and Baskakov [5]. The best possible result (corresponding to Theorem 4) was given by Baskakov [5].

§ 5. The initial exposition of the method of harmonic analysis is in the articles of Zhikov [44] and [45]. Later it was explained how certain restrictions in these articles can be relaxed (Boles [12] and especially Baskakov [4]).

Equation (20) in the scalar case of constant A_k was studied by Bohr & Neugebauer [23]; the main result (the almost periodicity of every bounded solution) is a generalisation of the Bohl–Bohr theorem. An investigation of equation (20) in a Banach space was started by Zhikov [45] and completed by Boles & Tsend [14].

Proposition 4 in the scalar case of constant A_k is known as 'Esclangon's lemma' [122] (see Levitan [76], p. 186). Hadamard & Landau (see Hardy, Littlewood & Polya [117]) have given another approach and proof of Esclangon's lemma. The general idea of our proof of Proposition 4 is close to the approach of Hadamard and Landau, but it seems that the inequality (22) was not noted earlier.

7 Stability in the sense of Lyapunov and almost periodicity

Notation

We shall use the following terminology and notation:

The term *compact flow* means a flow defined on a compact metric space;

$(\mathscr{H}, t)$ denotes a compact minimal isometric flow;
(M, t) denotes a compact minimal flow;
X is a complete metric space;
$S(t)\ (t \geqslant 0)$ is a continuous semigroup on X.

1 The separation properties

1. A fixed trajectory $\mathring{x}(t)$ of a semigroup $S(t)$ is called *separated* (respectively, *semiseparated*) if

$$\inf_{t \in J} \rho(\mathring{x}(t), x(t)) > 0$$

$$(\text{respectively, } \varliminf_{t \to -\infty} \rho(\mathring{x}(t), x(t)) > 0) \qquad (*)$$

for every trajectory $x(t) \not\equiv \mathring{x}(t)$.

When every trajectory is separated (semiseparated) the semigroup and flow are called *distal* (*semidistal*).

The simplest example of a distal flow is a flow $(\mathscr{H}, t)$; another example is any compact equicontinuous flow. But there are examples of compact distal flows that are not equicontinuous (see 'Bohr's example' at the end of this section).

A semigroup $S(t)$ is called an *extension* of a minimal flow (M, t) if there is a continuous mapping $j: X \to M$ such that

$$j(S(t)x) = (j(x))^t \quad (x \in X,\ t \geqslant 0).$$

Since every semitrajectory is dense in M we have $j(X)=M$. The inverse image $j^{-1}(h)$ is called a *fibre* over $h \in M$ and is denoted by X_h. Obviously, the space is partitioned into fibres, and under the action of $S(t)$ fibres go into fibres.

We say that a trajectory $\mathring{x}(t)$ is *separated* (*semiseparated*) *in a fibre* if the inequality $(*)$ holds for every trajectory $x(t) \not\equiv \mathring{x}(t)$ such that $j(x(t)) = j(\mathring{x}(t))$.

A fibre is called *distal* (*semidistal*) if every trajectory passing through it is separated (semiseparated) in the fibre.

We give an example of an extension which is important for what follows, and also examples of separated trajectories.

2. We consider in a Banach space B a non-autonomous evolutionary equation

$$u'_t = F(t)u, \tag{1}$$

where F is in general a non-linear unbounded operator (the notation u'_t is used for partial differentiation $\partial u/\partial t$). For the time being the nature of the operator F is not essential because our analysis will be based only on the existence and properties of a solving operator. We assume that the dependence of F on t is almost periodic, that is, $F(t)$ is an almost periodic function with values in some metric space R. Therefore we can consider (at least formally) a family $\mathcal{H} = \mathcal{H}(F)$ of 'limit' operator-functions $h = \hat{F}(s)$ of the form $\hat{F}(s) = \lim_{m\to\infty} F(s + t_m)$. We denote a shift on the space $\mathcal{H}$ by h^t; we also denote the initial operator-function $F(s)$ by h_0 and identify it with (1). For every $h \in \mathcal{H}$ we introduce the corresponding 'limit' equation

$$u'_t = \hat{F}(t)u. \tag{1_h}$$

Suppose (for the time being, formally) that $S_h(t)p$ $(t \geq 0,\ S_h(0)p = p)$ denotes a solution of equation (1_h) corresponding to an initial condition $p \in B$. We set $X = B \times \mathcal{H}$ and define a transformation $S(t): X \to X$ $(t \geq 0)$ by

$$S(t)x = S(t)\{p, h\} = \{S_h(t)p, h^t\}.$$

It can be verified in a straight-forward way that the operators $S(t)$ $(t \geq 0)$ commute. If the following continuity condition holds, then these transformations form a continuous semigroup.

Continuity condition. For any initial value $p \in B$ and any $h \in H$, the equation (1_h) has a unique strongly continuous solution $u(t) = S_h(t)p$

($t \geq 0$, $u(0)=p$), and the mapping

$$S_h(t): B \times \mathcal{H} \to B \tag{2}$$

is continuous for every $t \geq 0$.

We call the semigroup $S(t)$ a *basic semigroup*; it is an extension of a dynamical system $(\mathcal{H}, t)$, and the corresponding homomorphism is the projection $j: X \to \mathcal{H}$. In this case a trajectory is a pair $\{\hat{u}(t), h^t\}$, where $\hat{u}(t)$ is the solution of (1_h) which is defined on the whole line.

Now we turn to a more detailed study of the trajectories of a basic semigroup.

Let $\mathring{p}(t)$ $(t \in J)$ be a fixed solution of the original equation (1). We say that it is *separated* (*semiseparated*) if

$$\inf_{t \in J} \|\mathring{p}(t) - p(t)\| > 0 \; (\varliminf_{t \to -\infty} \|\mathring{p}(t) - p(t)\| > 0)$$

for every solution $p(t) \not\equiv \mathring{p}(t)$. It is clear that we can associate with a separated (semiseparated) solution $\mathring{p}(t)$ a trajectory $\{\mathring{p}(t), h_0{}^t\}$, which is separated (semiseparated) in a fibre.

The semiseparation property holds in the important case when a solution $\mathring{p}(t)$ is uniformly stable in the sense of Lyapunov (we omit the simple verification of this fact). If $\mathring{p}(t)$ is compact as well, then we can consider the limits $\mathring{p}(t+t_m) \to_{\text{loc}} \hat{p}(t)$. It is important that $\hat{p}(t)$ is also a uniformly Lyapunov stable solution of (1_h). Next we observe that for a basic semigroup (and its invariant subsets), we only need to verify the separation property for trajectories from one fibre. Indeed, the separation of the trajectories $\{p_1(t), h_1{}^t\}$, $\{p_2(t), h_2{}^t\}$ from different fibres (that is, $h_1 \neq h_2$) is ensured automatically by the distal property of the system $(\mathcal{H}, t)$. Hence we obtain the following result.

Proposition 1. *Let $\mathring{p}(t)$ be a compact uniformly Lyapunov stable solution. Then the closure of the corresponding trajectory $\{\mathring{p}(t), h_0{}^t\}$ is a semidistal set.*

We end this section by giving a classical example of Bohr which is important in a number of situations.

Example 1. Let $a(t)$ be a real almost periodic function whose indefinite integral $g(t) = \int_0^t a(s)\,ds$ cannot be represented as $g(t) = c_0 t + \phi(t)$, where c_0 is a constant and $\phi(t)$ is an almost periodic function.

We consider the non-autonomous equation $p'_t = ia(t)p$ and the corresponding dynamical system on $X = R^2 \times \mathcal{H}$ (R^m denotes the

m-dimensional Euclidean space). It is obvious that this system is distal, since solutions of the linear equation $p'_t = i\hat{a}(t)p$ satisfy the identity $\|p(t)\| = \|p(0)\|$. On the other hand, the non-trivial solutions are not almost periodic; this follows from Bohr's theorem of the argument. Thus, we have a distal but not uniformly continuous flow. By refining the properties of $a(t)$ we can obtain sharper negative results, for instance, a minimal distal but not strictly ergodic[1] flow on a three-dimensional torus. Bohr's example will be used significantly in Chapter 8, § 7, where, in particular, we give a 'concrete' construction of $a(t)$.

2 A lemma about separation

1. We denote by X^X the set of all mappings $X \to X$ endowed with the topology of pointwise convergence (the Tikhonov topology). The space X^X is a semigroup with respect to the composition uv (by definition, $(uv)(x) = u(v(x))$). It follows directly from the definition of pointwise convergence that the composition uv is continuous in u for any $v \in X^X$, and continuous in v if u is a continuous mapping. From here we obtain the following simple fact.

Proposition 2. *Let $A \subset X^X$ be a semigroup of continuous operators. Then the closure $\bar{A}$ is also a semigroup. If A is commutative, then the elements of A commute with those of $\bar{A}$.*

Proof. Let $u_\alpha \to u_0$ and $v_\beta \to v_0$. From the continuity of the composition uv with respect to an argument u it follows that $u_\alpha v_0 \in \bar{A}$. But then $u_\alpha v_0 \to u_0 v_0 \in \bar{A}$, proving the first part of the proposition. Next if $uv_\alpha = v_\alpha u$ $(u, v_\alpha \in A)$, then by taking the limit we obtain $uv_0 = v_0 u$ $(u \in A, v_0 \in \bar{A})$. We observe that the commutativity of the semigroup A does not by any means imply that of $\bar{A}$. Proposition 2 is proved.

Now we suppose that a dynamical system $S(t)$ acts on the space X; we denote the closure of the set $\{S(t)\}$ $(t \in J)$ in X^X by $T = T(X)$, and call it the *enveloping Ellis semigroup*. There is defined on the space T a natural dynamical system $\pi^t : \pi^t u = u^t = S(t)u = uS(t)$. For the concept of an Ellis semigroup to be meaningful we must assume, in addition, that all the trajectories of the system $S(t)$ are compact. In this case T is a compact Hausdorff space.

[1] A flow is called *strictly ergodic* if it has a unique invariant measure.

2. Suppose that a continuous semigroup $S(t)$ defined on X is an extension of a minimal system (M, t).

Lemma 1 (*The separation lemma*). *The following assertions hold:*

(1) *a compact trajectory that is semiseparated in a fibre is recurrent;*

(2) *two compact trajectories $x_1(t)$ and $x_2(t)$ from a single fibre, each of which is semiseparated in the fibre, are jointly recurrent and mutually separated, that is,* $\inf_{t\in J} \rho(x_1(t), x_2(t))>0$.

Proof. First of all we assume that there is defined on X a group and not a semigroup. Since we are concerned with the properties of compact trajectories, we can assume from the outset that X is compact.

We denote the fibre over an element $h \in M$ by $\Gamma(h)$, and the closure of the family $\{S(t),\ t\leqslant 0\}$ in X^X by T^-; T^- is a semigroup with respect to composition that is invariant under the transformations π^t $(t\leqslant 0)$.

The set $\bigcap_{t\leqslant 0} \pi^t T^-$ is non-empty and invariant under the group of transformations π^t $(t\in J)$. By Birkhoff's theorem, there exists a compact minimal set $V\subset T^-$. The set V is also a semigroup with respect to composition. Indeed, it follows from minimality that $V=\overline{\{\pi^t v_0\}}=\overline{\{S(t)v_0\}}$, where v_0 is an element from V. Then it is enough to refer to Proposition 2.

We introduce a family of continuous mappings $\phi_x : V\to X$ defined by

$$\phi_x(v)=v(x)\quad (v\in V, x\in X).$$

It follows easily from the minimality of V that $\phi_x(V)$ is a minimal set in X for every $x\in X$.

Hence it follows that the trajectory $x_0{}^t$ is recurrent provided that we can prove that $x_0\in\phi_{x_0}(V)$. We set $h_0=j(x_0)$.

The group π^t acting on V is an extension of the system (M, t). To find the corresponding homomorphism $l: V\to M$ we must fix an element $x\in X$ and set

$$l: V\xrightarrow{\phi_x} X\xrightarrow{J} M.$$

If $l^{-1}(h_0)$ is the fibre over h_0, then the transformations $v\in l^{-1}(h_0)$ have the property $v(\Gamma(h_0))\subset\Gamma(h_0)$. We denote by V_{h_0} the set of restrictions of these transformations to $\Gamma(h_0)$, and by δ the restriction operation. Since δ is continuous, V_{h_0} is a semigroup of transformations $\Gamma(h_0)\to\Gamma(h_0)$. Next we use the following important fact from the theory of topological semigroups.

Proposition 3. *Let V be a compact Hausdorff space with the structure of a semigroup, where the operation $\alpha\to\alpha\beta$ is continuous for any $\beta\in V$. Then V contains an idempotent element.*

We apply this proposition to the semigroup V_{h_0}. Suppose that ω is an idempotent element, that is, $\omega^2=\omega$, and let $\omega=\delta(v)$. We are going to prove that $v(x_0)=x_0$.

Since $v\in T^-$ there is a generalised sequence $S(t_\lambda)$ $(t_\lambda\leq 0)$ for which

$$S(t_\lambda)\xrightarrow{X^X} v.$$

By setting $x_1=v(x_0)$ we have

$$\begin{aligned} v(x_0)&=\lim S(t_\lambda)x_0=v^2(x_0)\\ &=v(x_1)=\lim S(t_\lambda)x_1. \end{aligned}$$

The relation $\lim S(t_\lambda)x_0=\lim S(t_\lambda)x_1$ does not contradict the semiseparation of the trajectory $x_0{}^t$ only in the case when $v(x_0)=x_1=x_0$. Thus, $v(x_0)=x_0$, and therefore x_0 belongs to the minimal set $\phi_{x_0}(V)$. This proves assertion (1) of Lemma 1.

To prove assertion (2) we consider the natural dynamical system on the space

$$Z=\{\{x,y\}\in X\times X: j(x)=j(y)\}. \tag{3}$$

Since the trajectory $\{x_1{}^t, x_2{}^t\}$ is semiseparated in a fibre it is recurrent (by assertion (1)). Since the metric ρ is continuous on Z, the function $\rho(x_1{}^t, x_2{}^t)$ is recurrent (see Chapter 1, § 6). Therefore

$$\inf_{t\in J}\rho(x_1{}^t, x_2{}^t)=\inf_{t\leq 0}\rho(x_1{}^t, x_2{}^t),$$

which proves Lemma 1 in the case of a group of transformations.

To prove Lemma 1 in the case of a semigroup we need quite a little more.

Let $\Phi(X)$ be a collection of continuous functions $J\to X$ with the topology of uniform convergence on every finite interval, and let γ be the metric corresponding to this convergence.

There is defined on the product $\Phi(X)\times M$ an obvious dynamical system: if $g\in\{f(s), h\}$ then $g^t=\{f(t+s), h^t\}$.

We consider two elements $g_1=\{x_1(s), h_0\}$ and $g_2=\{x_2(s), h_0\}$, where $x_1(s)$ and $x_2(s)$ are two trajectories of the original semigroup. It follows from Proposition 2 in Chapter 1 that the trajectories $g_1{}^t$ and $g_2{}^t$ are compact. Let G_0 be the smallest compact invariant subset containing these two trajectories. The corresponding fibre $\Gamma(h_0)$ consists of elements of the form $g = \{x(s), h_0\}$, where $x(s)$ is a trajectory of the original semigroup and $j(x(0)) = h_0$.

From the assumption that the trajectories $x_i(t)$ $(i = 1, 2)$ are semiseparated in a fibre, it follows immediately that the trajectories g_i^t are semiseparated in a fibre, that is,

$$\inf_{t \leq 0} \gamma(x(t+s), x_i(t+s)) > 0 \quad (x(s) \not\equiv x_i(s)).$$

Therefore, Lemma 1 for a semigroup follows from Lemma 1 for dynamical systems, and so this lemma is proved.

Now we prove Proposition 3.

Let Ω be the class of all non-empty subsets $A \subset V$ such that $AA \subset A$. Then $\Omega \neq \varnothing$ since $V \in \Omega$. We order Ω by inclusion, then by Zorn's lemma it contains a minimal element B. If $\omega \in B$, then $B\omega$ is compact, and $(B\omega)(B\omega) \subset BB\omega \subset B\omega$; consequently, $B\omega \in \Omega$ and $B\omega \subset B$. Since B is minimal, $B\omega = B$. Therefore there exists an element $p \in B$ such that $p\omega = \omega$.

Let $L = \{a \in B : a\omega = \omega\}$; then $p \in L$. Since a multiplication of an element on the right is continuous in V, the set L is closed and therefore compact. If $k, l \in L$, then $l(k\omega) = l\omega = \omega$, that is, $LL \subset L$. Thus, $L \in \Omega$ and $L \in B$, that is, $L = B$. Hence $\omega \in L$, that is, $\omega^2 = \omega$, and Proposition 3 is proved.

3. Let (X, t) be a flow with compact trajectories which is an extension of a minimal flow (M, t), $T = T(X)$ be an Ellis semigroup, and h_0 be a fixed element from M. We consider the set of those transformations $u \in T$ for which $u(X_{h_0}) \subset X_{h_0}$, and let T_{h_0} be the set of restrictions of these transformations to X_{h_0}.

Lemma 2. *If the fibre X_{h_0} is distal, then T_{h_0} is a group.*

Proof. We choose an arbitrary element $\beta \in T_{h_0}$ and are going to prove that $\beta^{-1} \in T_{h_0}$. The set $A = T_{h_0}\beta$ is a compact semigroup, and so it contains an idempotent ω. By using the same arguments as in Lemma 1, we see that $\omega(x) = x$ $(x \in X_{h_0})$. Thus, $\alpha_1\beta = e$ (the unit element of T_{h_0}) for some $\alpha_1 \in T_{h_0}$. Similarly, we can find an $\alpha_2 \in T_{h_0}$ such that $\alpha_1\alpha_2 = e$. Hence $\beta = e\beta = \alpha_2\alpha_1\beta = \alpha_2$, that is, $\alpha_1 = \beta^{-1}$, thus proving Lemma 2.

We consider the special case of Lemma 2 when the space M is a point. We obtain that the Ellis semigroup of a distal flow is a group (Ellis's theorem).

We introduce in the set of fibres the Hausdorff metric

$$\rho(X_{h_1}, X_{h_2}) = \sup_{x_2 \in X_{h2}} d(X_{h_1}, x_2) + \sup_{x_1 \in X_{h1}} d(x_1, X_{h_2}).$$

Then the following proposition holds.

Proposition 4. *If a fibre X_{h_0} is distal, then the mapping $h \to X_h$ is continuous at h_0.*

Proof. Let $h_m \to h_0$ and $x_0 \in X_{h_0}$. We require to prove that we can find elements $x_m \in X_{h_m}$ for which $\lim_{m\to\infty} x_m = x_0$. Let $u_m \in T$ be any element such that $u_m(X_{h_0}) \subset X_{h_m}$ (it follows from the minimality of M such that such a u_m exists). The set $\{u_m\}$ has a limit point $\alpha \in T$. Since $\alpha \in T_{h_0}$, by Lemma 2 $\alpha(X_{h_0}) = X_{h_0}$, that is, $\alpha(x_1) = x_0$ for $x_1 \in X_{h_1}$. But then $x_0 = \alpha(x_1) = \lim u_m(x_1)$, which proves our proposition.

Note that the continuity of the mapping $h \to X_h$ at each point means that the mapping $j: X \to M$ is open.

3 Corollaries of the separation lemma

Below we suppose that X is a complete metric space with an invariantly acting semigroup $S(t)$ $(t \geq 0)$. It is assumed that every trajectory of this semigroup is compact.

1. The first group of corollaries is obtained by applying Lemma 1 in the special case when M is a point.

Corollary 1. *If X is compact and the semigroup $S(t)$ is semidistal, then $S(t)$ is a distal flow.*

Proof. Because the distal property of $S(t)$ is obtained at once from Lemma 1, we prove that $S(t)$ is a group. Since all the trajectories are recurrent, X is split into minimal sets. If $\mathring{X}$ is minimal, then $S(t)\mathring{X} = \mathring{X}$ for every $t \geq 0$. Hence it follows that $S(t)X = X$ for every $t \geq 0$, and consequently, the inverse transformations $S^{-1}(t)$ are continuous, as we required to prove.

***Remark* 1.** If the space X is not compact, then in general, our semidistal semigroup is not a flow since in general the operators $S^{-1}(t)$ $(t \geq 0)$ are not continuous. The continuity of these operators is easily ensured by the following additional condition: for every compact set $K \subset X$ the set

$$\overline{\bigcup_{t\geq 0} S(t)K} \text{ is compact.} \tag{4}$$

The following condition for a trajectory to be absolutely recurrent holds.

Corollary 2. *A compact trajectory x_0^t is absolutely recurrent if and only if it is semiseparated in $X_0=\overline{\{x_0^t\}}$, that is, it is semiseparated in its closure.*

Proof. The necessity of the condition is obvious. Indeed, if a trajectory x_0^t is absolutely recurrent, then the pair $\{x_0^t, x^t\}$ is recurrent for any trajectory x^t from X_0. If $x^t \not\equiv x_0^t$, then the minimal set $\{x_0^t, x^t\}$ must be separated from the diagonal set in $X_0 \times X_0$, which means that the trajectories x_0^t and x^t are mutually separated, that is, $\inf_{t\in J} \rho(x_0(t), x(t))>0$.

To prove the sufficiency we consider an arbitrary recurrent trajectory y_0^t (of some semigroup on Y). Let

$$\mathring{Z}=\overline{\{x^t, y_0^t\}}, \quad \mathring{Y}=\overline{\{y_0^t\}},$$

and $j:\mathring{Z}\to Y$ be the projection onto the second component. Consider the fibre $j^{-1}(y_0)$. The trajectories passing through it have the form $\{x^t, y_0^t\}$. Hence it is clear that the trajectory $\{x_0^t, y^t\}$ is semiseparated in the fibre, and so by Lemma 1 it is recurrent, as we required to prove.

In particular, an almost periodic trajectory is absolutely recurrent.

2. We say that an extension $j:X\to M$ is *positively stable* if for every $\varepsilon>0$ and every compact set $K\subset X$ there is a $\delta=\delta(\varepsilon; K)$ such that

$$\rho(S(t)x_1, S(t)x_2)\leqslant \varepsilon \quad (t\geqslant 0)$$

whenever $\rho(x_1, x_2)\leqslant\delta$ and $j(x_1)=j(x_2)$, $x_1, x_2\in K$.

It is useful to note that from the positive stability of an extension it follows at once that every fibre is semidistal. Therefore, from Lemma 1 we have

Corollary 3. *A semigroup $S(t)$ is a distal and two-sidedly stable extension. When X is compact or condition* (4) *holds, then $S(t)$ is a flow.*

In the proof we need to prove only the two-sided stability property. For the proof we consider on the set Z defined by (3) the continuous function $p(z)=\rho(x, y)$. Since all the trajectories z^t are recurrent, the function $p(z^t)$ is recurrent. Therefore

$$\sup_{t\in J} \rho(x^t, y^t)=\sup_{t\geqslant 0} \rho(x^t, y^t) \quad (j(x)=j(y)),$$

which gives the two-sided stability of the extension.

4 Corollaries of the separation lemma (continued)

1. We return to the properties of compact solutions of equation (1), and recall that we mean solutions defined on the whole time axis.

Corollary 4. *Let $p_1(t)$ and $p_2(t)$ be semiseparated compact solutions. Then they are jointly recurrent and mutually separated, that is*

$$\inf_{t\in J} \|p_1(t)-p_2(t)\|>0. \tag{5}$$

Corollary 5. *Let $\mathring{p}(t)$ be a uniformly Lyapunov stable solution and $\mathring{X}$ be the closure of the corresponding trajectory. Then the restriction of the basic semigroup to $\mathring{X}$ is a minimal distal flow.*

For the proof we need to take into account Proposition 1 and Corollary 1.

Next, let $V(p, q)$ be a non-negative continuous function on $B\times B$ which is zero only on the diagonal.

We say that equation (1) is *V-monotonic* if for every two solutions $p_1(t)$, $p_2(t)$ of (1) the scalar function $V(p_1(t), p_2(t))$ is non-increasing on that part of the time axis where both solutions are defined.

Corollary 6. *If* (1) *is V-monotonic for any pair of compact solutions $p_1(t)$ and $p_2(t)$ $(t\in J)$, then the 'identical' invariance $V(p_1(t), p_2(t))\equiv$ constant holds.*

Indeed, the solutions $p_1(t)$, $p_2(t)$ are jointly recurrent, and therefore the function $g(t)=V(p_1(t), p_2(t))$ is recurrent. Since (1) is V-monotonic $g(t)$ is non-increasing. But then it is recurrent only when it is identically constant.

The properties of mutual separation and identical invariance play an especially important role in what follows. We show that these properties are peculiar to equations with almost periodic coefficients and do not hold, for instance, in the case of bounded coefficients.

In Lemma 1 the only condition on the system (M, t) is minimality. This means that the corresponding corollaries (in particular, the separation property (5) and identical invariance) are valid for equations with recurrent coefficients. But it is impossible to weaken recurrence to boundedness, as is shown by the scalar equation $p'_t+a(t)p(t)=0$, where $a(t)$ is a continuous function such that $a(t)=0$ for $t\leqslant 0$ and $a(t)=1$ for $t\geqslant 1$. Here all the solutions are bounded and semiseparated but not mutually separated. Our monotonicity condition holds ($V=|p-q|$) but identical invariance does not.

Let us also note that the two-sided uniqueness theorem holds in the class of semiseparated solutions; this assertion is not true without the recurrence condition, as is not difficult to demonstrate by an example.

2. We denote by $U(t, \tau)$ $(t \geq \tau)$ an operator $B \to B$ which maps an initial value $p = p(\tau) \in B$ into the value of the solution of (1) at the point t.

Suppose that uniform positive stability holds in the sense that the operators $U(t, \tau)$ $(t \geq \tau)$ are equicontinuous on each compact set in B.

In our earlier notation we have $U(t, \tau) = S_{h_0}\tau(t - \tau)$. Since the set $\{h_0{}^\tau\}$ is dense in $\mathcal{H}$ and the reflection (2) is continuous, the family of operators

$$S_h(t): B \to B \quad (t \geq 0, h \in \mathcal{H}) \tag{6}$$

is equicontinuous on compact sets, that is, the semigroup $S(t)$ on $B \times \mathcal{H}$ is a positively stable extension of the flow $(\mathcal{H}, t)$. Hence, if X is the collection of all compact trajectories in $B \times \mathcal{H}$, then it is a distal two-sidedly stable extension (see Corollary 3).

We discuss especially the case when B is finite-dimensional. Then the set X is closed in $B \times \mathcal{H}$. For suppose that $x_m = \{p_m, h_m\} \in X$, $x_m(t) = \{p_m(t), h_m{}^t\}$ are the corresponding compact trajectories, and $x_m \to \{\hat{p}, \hat{h}\}$. By hypothesis the operators (6) are equicontinuous on compact sets; therefore, these operators are uniformly bounded on compact sets provided that $X \neq \varnothing$. From here and from the recurrence property of the solutions $p_m(t)$ we have

$$\sup_{t \in J, m} \|p_m(t)\| = \sup_{t \geq 0, m} \|p_m(t)\|$$

$$= \sup_{t \geq 0, m} \|S_{h_m}(t) p_m(0)\| < \infty.$$

By taking the limit we see that the solution $\hat{p}(t)$ is bounded, that is, X is closed. Then condition (4) follows easily from the uniform boundedness of the operators (6). Thus, the following assertion holds.

Corollary 7. *Let B be finite-dimensional and suppose that the uniform positive stability condition holds. Then the set of compact trajectories is closed in $B \times \mathcal{H}$, and the restriction of the basic semigroup to this set is a distal flow which is a two-sidedly stable extension of the flow $(\mathcal{H}, t)$.*

5 A theorem about almost periodic trajectories

1. In this section we are going to discuss the simplest principles that enable us to judge whether almost periodic trajectories or solutions exist.

Let (X, t) be a compact flow which is an extension of a flow $(\mathcal{H}, t)$, $j: X \to \mathcal{H}$ be the corresponding homomorphism, and $\Gamma(h)$ be the fibre over an element h.

Theorem 1 (*Favard* [106]). *If each fibre $\Gamma(h)$ consists of a single element, then all the trajectories are almost periodic.*

To prove the theorem it is enough to note that in this case the homomorphism $j: X \to \mathcal{H}$ is an isomorphism, that is, the inverse mapping $j^{-1}: \mathcal{H} \to X$ is continuous.

Theorem 2 (*Levitan* [74]). *If a fibre $\Gamma(h_0)$ consists of a single element, then the corresponding trajectory is an N-almost periodic function.*

For a proof we consider the 'winding' $\{h_0^t\}$ and give it the topology of a subspace of $\mathcal{H}$. The correspondence $h_0^t \to x(t)$, where $j(x(0)) = h_0$, defines a function on the winding. From the fact that every fibre $\Gamma(h_0^t)$ consists of one point and the set $X = \bigcup_{h \in \mathcal{H}} \Gamma(h)$ is closed, it follows that our function is continuous on the winding. But then $x(t)$ is N-almost periodic.

***Remark* 2.** In applications, it is usual to preassign on a space X (chosen in a suitable way) not a flow but an invariantly acting semigroup. Clearly, Theorem 1 still holds in this case; but in Theorem 2 we need to require that every fibre $\Gamma(h_0^t)$ $(t \in J)$ consists of a single element.

The next theorem is a fairly significant generalisation of Favard's theorem.

Theorem 3 (*Amerio* [1]). *If a flow (X, t) is distal and some fibre $\Gamma(h_0)$ is finite, then all the trajectories are almost periodic.*

Proof. We show that all the fibres are finite and consist of one and the same number m of elements. For suppose that $\Gamma(h_0) = \{x_1, \ldots, x_m\}$, $\{y_1, y_2, \ldots, y_k\} \subset \Gamma(h_1)$, and $k > m$. We choose a sequence $\{t_n\} \subset J$ for which $\lim_{n\to\infty} h_1^{t_n} = h$ and $\lim_{n\to\infty} y_i(t_n) = z_i$ exist $(1 \leq i \leq k)$. Then $z_i \in \Gamma(h)$, and $z_i \neq z_j$ $(i \neq j)$ because the fibre $\Gamma(h_1)$ is distal; this contradicts the inequality $k > m$.

Let $\Delta(h) = \inf_{x \neq y \in \Gamma(h)} d(x^t, y^t)$; similar arguments show that $\Delta(h) \equiv \Delta > 0$.

Suppose that h is an arbitrary element in $\mathcal{H}$. It is not difficult to see that there is a neighbourhood $\Omega=\Omega(h)$ of h such that the set $\Gamma(\Omega)=\bigcup_{h\in\Omega}\Gamma(h)$ is partitioned into m parts $\Pi_1, \Pi_2, \ldots, \Pi_m$ and that the following conditions are satisfied: (1) $d(\Pi_i, \Pi_j)\geq\Delta/2$ $(i\neq j)$; (2) if $h\in\Omega$, then $\Gamma(h)=\{\phi_i(h)\}$, where $\phi_i(h)$ are continuous on Ω.

By selecting a finite covering of the space $\mathcal{H}$ by neighbourhoods $\Omega(h)$ we obtain that X is an m-sheeted covering of $\mathcal{H}$. Therefore, for any $\varepsilon\in(0, \Delta/2)$ we can find a $\delta=\delta(\varepsilon)$ so that one of the following inequalities holds:

$$\rho(x_1, x_2)\leq\varepsilon, \quad \rho(x_1, x_2)\geq\Delta/2, \tag{7}$$

provided that $x_1\in\Gamma(h_1)$, $x_2\in\Gamma(h_2)$ and $d(h_1, h_2)\leq\delta(\varepsilon)$.

Now we are going to prove that the flow (X, t) is uniformly continuous. Suppose that $\rho(x_1, x_2)\leq\varepsilon\rho(h_1, h_2)\leq\delta(\varepsilon)$, $j(x_1)=h_1$, and $j(x_2)=h_2$. We shall show that $g(t)=\rho(x_1{}^t, x_2{}^t)\leq\varepsilon$ for any $t\in J$. Indeed, if we assume that $g(t_1)>\varepsilon$ (taking for definiteness $t_1>0$), then since $\rho(h_1{}^t, h_2{}^t)=\rho(h_1, h_2)\leq\delta/\varepsilon)$ by (7) we have $g(t_1)\geq\Delta/4$. Since $g(0)\leq\varepsilon\leq\Delta/4$ and $g(t_1)\geq\Delta/4$, we can find a $t_0\in(0, t_1)$ for which $g(t_0)=\rho(x_1{}^{t_0}, x_2{}^{t_0})=\Delta/3$. But because $\rho(h_1{}^{t_0}, h_2{}^{t_0})\leq\delta(\varepsilon)$ we obtain a contradiction to (7), and the theorem is proved.[2]

Levitan's theorem, in contrast to those of Favard and Amerio, operates with a fixed fibre. This gives rise to the important question of the possibility of generalising Levitan's theorem in the spirit of Amerio's result. In other words, are the trajectories of a distal fibre N-almost periodic functions if the fibre is finite? Unfortunately, the answer to this question turns out to be negative; a counterexample is given at the end of this section.

Another question, which arises in connection with Amerio's theorem, is whether the finiteness condition can be lifted. This question is resolved by the following theorem.

Theorem 4 (*Zhikov* [50]). *If a flow* (X, t) *is distal and if some fibre has dimension* 0, *then all the trajectories are almost periodic.*

A proof is given in § 6.

Theorems 1–4 form a system of 'elementary principles' to which in the final analysis all theorems about almost periodic solutions reduce. Later on we shall develop different methods that will enable

[2] It may seem that the last step relies on the connectedness of the real line. However, we do not use it here, and the proof we have given is suitable without changes for discrete dynamical systems (cascades).

us to find for the basic semigroup compact invariant sets with appropriate properties; but now we dwell only on a simple application of Amerio's theorem.

Proposition 5. *Let* $\mathring{p}(t)$ *be a compact uniformly Lyapunov stable solution of* (1), *and assume that* $\mathring{p}(t)$ *is asymptotically stable. Then this solution is almost periodic.*

Proof. By Corollary 4 the set $\mathring{X}=\overline{\{\mathring{p}(t), h_0{}^t\}}$ is distal and minimal. The fibre $\Gamma(h_0)$ contains the point $\mathring{p}(0)$; it follows immediately from the asymptotic stability of $\mathring{p}(t)$ and the distal character of $\Gamma(h_0)$ that $\mathring{p}(0)$ is an isolated point in $\Gamma(h_0)$. But then it is easy to obtain from considerations of minimality and distalness that every point of the fibre $\Gamma(h_0)$ is isolated, that is, $\Gamma(h_0)$ is finite. Now we only need to use Theorem 3.

2. Example 2. Now we are going to construct a counterexample to show that it is impossible to generalise Theorem 2 in the spirit of Amerio's theorem. The construction is based on the following auxiliary proposition.

Let $f(t)$ be (in general) a complex valued almost periodic function satisfying the condition

$$|f(t)|>0 \quad (t\in J), \quad \inf_{t\in J}|f(t)|=0.$$

In addition, we assume that every function $\hat{f}(t)\in\mathscr{H}(f)$ has not more than one zero.

Proposition 6. *If the continuous root* $|f(t)|^{1/2}$ *is an* N*-almost periodic function, then it is almost periodic.*

Proof. We can write the N-almost periodic function $|f(t)|^{1/2}$ in the form $|f(t)|^{1/2}=v(h_0{}^t)$, where $h_0{}^t$ is the trajectory of some minimal isometric flow (G, t), and v is a continuous function on $\{h_0{}^t\}\subset G$.

The function v^2, being uniformly continuous on $\{h_0{}^t\}$, can be extended to a continuous function $f(h)$ on G (so that $f(t)=f(h_0{}^t)$).

We consider a sufficiently small neighbourhood Λ of h that

$$|f(h)|\geq\Delta>0 \quad (h\in\Lambda). \tag{8}$$

(This is possible because $f(h_0)=f(0)\neq 0$.)

Since v is continuous on $\{h_0{}^t\}$ we can find a neighbourhood $\Lambda_1\subset\Lambda$ such that

$$|u(h_0{}^t)-u(h_0)|\leq\Delta/2$$

for $h_0^t \in \Lambda_1$. It follows from (8) that the values of $v(h_0^t)$ for $h_0^t \in \Lambda_1$ are the same as those of a branch of $[f(h)]^{1/2}$ $(h \in \Lambda_1)$. In other words, $v(h_0^t)$ is extended to a function $v(h)$ continuous on Λ_1.

Now we prove that $v(h_0^t)$ can be extended to a function $v(h)$ continuous on G. Assuming otherwise we obtain a sequence $\{t_m\}$ and two subsequences $\{t'_m\}$, $\{t''_m\}$ such that

$$\lim h_0^{t_m} = h, \quad \lim [f(t'_m)]^{1/2} \neq \lim [f(t''_m)]^{1/2},$$
$$f(h) \neq 0.$$

The function $f(h^t)$ vanishes at not more than one point; to be specific suppose that $f(h^t) \neq 0$ for $t \geqslant 0$.

We can find a number $t_1 > 0$ such that $h^{t_1} = h_1 \in \Lambda_1$. Since $u(h)$ is continuous on Λ_1,

$$\lim [f(t_1 + t'_m)]^{1/2} = \lim [f(t_1 + t''_m)]^{1/2}.$$

It is now clear from continuity considerations that $\lim [f(t'_m)]^{1/2} = \lim [f(t''_m)]^{1/2}$, because $|f(t+t_m)| \geqslant \varepsilon > 0$ on $0 \leqslant t \leqslant t_1$. This proves Proposition 6.

Now we construct the counterexample. Let $\mathscr{H}$ be the two-dimensional torus with the minimal shift. We denote the trajectory $\{h^t\}$ by Jh, and fix three points h_0, h_1, $h_2 \in \mathscr{H}$ such that

$$h_1, h_2 \notin Jh_0, Jh_1 \neq Jh_2.$$

We consider a continuous complex-valued function $f(h)$ on $\mathscr{H}$ with the properties:

(1) $f(h_1) = f(h_2) = 0$, $f(h) \neq 0$ for $h \neq h_1$ and $h \neq h_2$;

(2) the index of the vector field $f(h)$ at h_1 is equal to 1 and at h_2 to -1.

Let $u_1(t) = [f(h_0^t)]^{1/2}$ and $u_2(t) = -u_1(t)$ be two continuous roots. We denote by X the closure of the trajectory $x_1^t = \{u_1(t+s), h_0^t\}$ in the space $\Phi(R^2) \times \mathscr{H}$. The fibre $\Gamma(h_0)$ consists of not more than two elements $x_1 = \{u_1(s), h_0\}$ and $x_2 = \{u_2(s), h_0\}$ (it will be clear later on that the case of a single element is excluded). It is important to note that the fibre $\Gamma(h_0)$ is distal, and hence the system (X, t) is minimal (by Lemma 1). If, for instance, x^t is an N-almost periodic trajectory, then $u_1(s)$ is an N-almost periodic function. Then (by Proposition 6) $u_1(s)$ is an almost periodic function. This means that the trajectories of the system (X, t) are almost periodic and each fibre consists of two elements. In this situation (as in general in the situation of Amerio's theorem), X is a finite-sheeted covering of the space $\mathscr{H}$. Therefore, we can find a neighbourood Λ of h_1 not containing h_2

such that

$$\Gamma(h)=\{\phi_1(h), \phi_2(h), h\},$$

where the $\phi_i=\phi_i(h, s)$ are continuous functions $\Lambda \to \Phi(R^2)$. But then the numerical functions $\phi_i(h, 0)=|f(h)|^{1/2}$ are continuous on Λ, which is impossible by conditions (1), (2). This completes the construction of the example.

In this example a distal fibre consists of two elements, but the corresponding trajectories are not N-almost periodic. The number of non-distal fibres in the example is two (with accuracy up to a shift in $\mathscr{H}$). Therefore, arguments that the 'majority' of fibres must be distal do not help us. A more detailed analysis shows that the constructed flow (X, t) is minimal, strictly ergodic and has a mixed spectrum.

6 Proof of the theorem about a zero-dimensional fibre

We split the proof of Theorem 4 into two separate lemmas.

Lemma 3. *If the extension $X\to\mathscr{H}$ is stable and some fibre $\Gamma(h_0)$ is zero-dimensional, then all the trajectories of the system (X, t) are almost periodic.*

Proof. Since we are speaking of a separate trajectory, we may assume that the system (X, t) is minimal. We set

$$Z=\{(x, y)\in X\times X: j(x)=j(y)\}. \tag{9}$$

We define on the set Z an invariant function $p(z)=p(x_1, x_2)=\sup_{t\in J}\rho(x_1^t, x_2^t)$. From the stability of the extension it follows that the metric p is equivalent to the original metric on every fibre. We fix a point $\mathring{x}\in\Gamma(h_0)$. The numerical set $A=\{p(\mathring{x}, x): x\in\Gamma(h_0)\}$ does not entirely contain any interval $[0, \varepsilon]$ because p is a metric on a zero-dimensional fibre. Then since functions of the form $\rho(x_1^t, x_2^t)$ are recurrent and the system (X, t) is minimal, we obtain $A=\{p(z): z\in Z\}$.

We prove that any trajectory x^t is almost periodic. By assuming that Bochner's criterion does not hold we have two sequences $\{t_m\}$, $\{\tau_m\}\in J$ such that $x^{t+t_m}\to_{\text{loc}}\hat{x}^t$, $\rho(x^{t_m+\tau_m}, \hat{x}^{\tau_m})\geqslant\delta>0$. Here it is clear that the convergence $j(x^{t+t_m})\to j(\hat{x}^t)$ is uniform on J, so that the trajectory $h^t=j(x^t)$ is almost periodic. Now we choose a number $\lambda\notin A$ with $0<\lambda\leqslant\delta$, and set $\phi_m(t)=\rho(x^{t+t_m}, \hat{x}^t)$. To be definite we assume that $\tau_m\geqslant 0$. Since $\phi_m(0)\to 0$ and $\phi_m(\tau_m)\geqslant\delta$ we can find points

$\theta_m \in [0, \tau_m]$ such that

$$\phi(\theta_m) = \lambda, \phi(t) \leq \lambda \quad \text{for } t \in [0, \theta_m]. \tag{10}$$

Obviously, $\theta_m \to +\infty$. By taking subsequences if necessary, we may assume the convergence: $x^{t+t_m+\theta_m} \to_{\text{loc}} y_1^t$ and $\hat{x}^{t+\theta_m} \to_{\text{loc}} y_2^t$. Clearly, $j(y_1) = j(y_2)$.[3] But from (10) we obtain

$$\sup_{t \leq 0} \rho(y_1^t, y_2^t) = \rho(y_1, y_2) = \lambda$$
$$= \sup_{t \in J} \rho(y_1^t, y_2^t) = p(y_1, y_2),$$

which contradicts the choice of λ. This proves Lemma 3.

Lemma 4. *Suppose that the extension $X \to M$ is distal. Then if some fibre is zero-dimensional, then this extension is stable.*

For the proof we need the concept of an equitransitive system (Gottschalk & Hedlund [38]).

A system (X, t) is called *equitransitive* at a point x if for any neighbourhood U of x we can find a neighbourhood V of x and a compact set $K \subset J$ such that $JV \subset KU$, where

$$JV = \bigcup_{t \in J} S(t)V, \quad KU = \bigcup_{t \in K} S(t)U.$$

Note that a transitive (that is, minimal) system is equitransitive. For if U is an open neighbourhood, then JU is open and invariant, that is, $JU = X$. But then, in view of compactness, it is enough to choose a finite covering.

The main step in proving Lemma 4 is the proof that the system (Z, t) is equitransitive at every point. The stability of the extension $X \to M$ follows fairly simply from equitransitivity. We give this reduction.

For a given open neighbourhood U of the diagonal in Z we choose an open neighbourhood W of the diagonal such that $\overline{W} \subset U$. By equitransitivity, for every $y \in Z \setminus U$ there exists an open neighbourhood Ω_y and a compact set $K_y \subset J$ such that $J\Omega_y \subset K_y(X \setminus \overline{W})$. Since the set $X \setminus U$ is compact, we can find a finite set $L \subset X \setminus U$ such that $X \setminus U = \bigcup_{y \in L} \Omega_y$. On setting $K = \bigcup_{y \in L} K_y$ we obtain $J(X \setminus U) \subset K(X \setminus W)$. The set $V = \bigcap_{t \in K} S(t)W$ is non-empty (it contains the diagonal), and open since the operators S_t, where $t \in K$, are equicon-

[3] Since $\lim_{m \to \infty} (\lim_{n \to \infty} h^{t+t_n+\theta_m} = \lim_{m \to \infty} (h^{t+t_m+\theta_m})$.

tinuous. Hence the relation we have obtained, rewritten in the form $JV \subset U$, gives the stability of the extension $X \to M$.

***Proof of Lemma* 4.** first of all let us note that from the zero-dimensionality of some fibre of a distal extension it follows trivially that every fibre is zero-dimensional.

For the natural extension $Z \to M$ we denote the fibre $\Gamma(h) \times \Gamma(h)$ by $E(h)$. For any set $\Lambda \subset M$ we put $E(\Lambda) = \bigcup_{h \in \Lambda} E(h)$. Let Λ be an open neighbourhood of $h = j(z)$ in M. Since the fibre $E(h)$ is zero-dimensional, for a sufficiently small diameter diam (Λ) we can find open sets P and Q such that

$$z \in P, \quad E(\Lambda) = P \cup Q, \quad \bar{P} \cap \bar{Q} = \varnothing. \tag{11}$$

Now we assume that the system (Z, t) is not equitransitive at z. Then the relation $JP \subset KP$ is not fulfilled for some neighbourhood P (of the form described above) and any compact set $K \subset J$. In particular, if we take $K = [-m, m]$, then we can find points $z_m \in P$ and $t_m \in J$ for which $z_m{}^t \notin P$ for $t \in [t_m - m, t_m + m] = \Delta_m$. Since $0 \notin \Delta_m$, to be definite we can assume that $\Delta_m \subset (0, \infty)$. We consider a maximal interval $[\alpha_m, \beta_m]$ such that

$$\Delta_m \subset [\alpha_m, \beta_m], \quad z_m{}^t \notin P \quad \text{for } t \in [\alpha_m, \beta_m]. \tag{12}$$

From the condition for maximality we obtain $z_m{}^{\alpha_m} \in \bar{P}$. Without loss of generality we may assume that $z_m{}^{\alpha_m} \to z_0$. Then from (11) we obtain immediately that $z_0 \in \bar{P}$ but $z_0{}^t \notin P$ for $t \geqslant 0$. In the recurrent case the closure of the trajectory coincides with the closure of a semitrajectory. Therefore,

$$z_0 \in \bar{P}, \quad \overline{\{z_0{}^t\}} \cap \bar{P} = \varnothing. \tag{13}$$

We write $h_0 = j(z_0)$ and $Z_0 = \overline{\{z_0{}^t\}}$. Since $h_0 \in \bar{\Lambda}$ we have $h_0 = \lim_{m \to \infty} h_m$, where $h_m \in \Lambda$. Because the mapping $j: Z_0 \to M$ is open (see Proposition 4), $z_0 = \lim_{m \to \infty} z_m$ for some sequence of points $z_m \in Z_0 \cap E(h_m)$. But from (13) and (11) it follows that $z_m \in Q$ and so $z_0 \in Q$. This contradicts the fact that $\rho(\bar{P}, \bar{Q}) > 0$ and the lemma is proved.

It remains to examine the numerical module corresponding to the almost periodic solution in Theorem 4. Let $\mathfrak{M}$ be the module corresponding to the almost periodic function $h_0{}^t$, and let $\mathfrak{M}_{\text{rat}}$ be its rational hull, that is, $\mathfrak{M}_{\text{rat}} = \{\lambda/m\}$, where $\lambda \in \mathfrak{M}$ and m is an integer. By using Kronecker's theorem it is not difficult to see that the module corresponding to the almost periodic solution in Theorem 4 belongs to $\mathfrak{M}_{\text{rat}}$.

7 Statement of the principle of the stationary point

Let the space $B = R^m$ be finite-dimensional. We assume that the uniform positive stability condition holds for (1), and that this equation has at least one solution bounded for $t \geqslant 0$. Let X be the collection of all the compact trajectories of the basic semigroup. Then $X \neq \emptyset$ and is closed in $B \times \mathcal{H}$. We denote the fibre X_{h_0} over h_0 by P, in other words, P is a set of initial values $\{p(0)\}$ corresponding to bounded solutions of (1). The set P has the following important property.

Proposition 7. *The set P is a homogeneous space relative to a compact topological transformation group; moreover, every element of this group of transformations is the restriction to P of a continuous mapping $R^m \to P$.*

Proof. We denote by T and T^+ the closure of the sets $\{S(t)\}$ and $\{S(t): t \geqslant 0\}$ in X^X, respectively. By Ellis's theorem, T is a group with respect to composition, and so the natural dynamical system on T is minimal. Hence we obtain the important relationship $T = T^+$.

We consider the set of elements $u \in T$ for which $u(\Gamma(h_0)) \subset \Gamma(h_0)$. Let $E = T_{h_0}$ stand for the set of restrictions of these transformations to $\Gamma(h_0) = P$; T_{h_0} is a group with respect to composition and is compact in the topology of pointwise convergence (see Lemma 2). From the condition for uniform positive stability and the equality $T = T^+$ it follows immediately that every $u \in T_{h_0}$ is the restriction to P of some continuous transformation $R^m \to P$. The set T_{h_0} is a topological group, since by the equicontinuity (on each compact set in P) of the transformations $u \in T_{h_0}$, the operation of composition is continuous in the collection of variables. This proves the proposition.

Now we introduce some more terminology.

Let A be a subgroup of the group E. By an *A-orbit* of $p \in P$ we mean the set $A_p = \{u(p): u \in A\}$. A set $P_0 \subset P$ is called *A-invariant* if it contains the A-orbit of each of its points. It is easy to see that P is partitioned into disjoint A-orbits, and that the group A acts transitively on each A-orbit.[4]

Since the trajectories of the system (X, t) are recurrent, the set X splits into minimal subsets. The fibre over an element h_0 corresponding to some minimal set is called a *minimal fibre*. It is sufficiently

[4] That is, for any two elements p_1, p_2 belonging to an A-orbit, there is a $u \in A$ such that $u(p_1) = p_2$.

clear that the partitioning of P into minimal fibres is the same as that into E-orbits. Thus, a minimal fibre is a homogeneous space with respect to a compact transitive transformation group. It can turn out that the group $E = T_{h_0}$ is not connected. Let G denote the component of the unit element. We say that a point $p_0 \in P$ is *stationary* if $u(p_0) = p_0$ for every $u \in G$. It is easily seen that if p_0 is stationary, then the minimal fibre containing it consists of stationary points. Indeed, the minimal fibre coincides with the E-orbit of p_0. Let $p_1 = u(p_0)$, where $u \in E$. Since $u^{-1}Gu = G$ (the component of the unit element is always a normal divisor), we have $G_{p_1} = \{p_1\}$.

Let $\mathfrak{M}$ stand for the numerical module corresponding to the right-hand side of (1), and $\mathfrak{M}_{\text{rat}}$ denote its rational hull.

Principle of the stationary point. *For an almost periodic solution with a module of Fourier exponents belonging to $\mathfrak{M}_{\text{rat}}$ to correspond to an initial value p_0 it is necessary and sufficient that p_0 is a stationary point.*

Proof. Assume that p_0 is a stationary point. By Theorem 4 it is enough to prove that the corresponding minimal fibre is zero-dimensional. Let $T_{h_0}^{p_0}$ denote the restriction of the group $E = T_{h_0}$ to the minimal fibre containing p_0. There is a natural homomorphism of T_{h_0} to $T_{h_0}^{p_0}$ (the restriction homomorphism). Clearly, under this homomorphism, the component of the unit element is mapped into the component of the unit element. Since the action of the group G on a minimal fibre is trivial, the component the unit element of the group $T_{h_0}^{p_0}$ is trivial, that is, the group $T_{h_0}^{p_0}$ is zero-dimensional. But the zero-dimensionality of a group implies the zero-dimensionality of the homogeneous space, that is, of the minimal fibre. The proof of the sufficiency does not present any difficulty.

Thus, at least formally, the problem about almost periodic solutions is reduced to one about a common fixed point of some compact topological transformation group. If the group G is non-commutative (in general this is the case), then the problem about a common fixed point is very difficult. A facilitation of this is the fact that the transformations forming G act on the entire space R^m and G is connected.

8 Realisation of the principle of the stationary point when the dimension $m \leqslant 3$

1. Suppose that the condition for uniform positive stability holds. Our aim in this section is to prove the following theorems.

Theorem 5. *Suppose that $m=2$ and that equation* (1) *has a bounded solution for $t \geqslant 0$. Then there exists at least one almost periodic solution with a module of Fourier exponents belonging to $\mathfrak{M}_{rat}$. Moreover, either the almost periodic solution with a module in $\mathfrak{M}_{rat}$ is unique, and so the module of this solution actually belongs to $\mathfrak{M}$, or every bounded solution is almost periodic and has a module in $\mathfrak{M}_{rat}$. In the first case, either the equation has a unique almost periodic solution, or every bounded solution is almost periodic.*

Let us note that all the possible cases considered in Theorem 5 are realised in Bohr's example. The trivial solution plays the role of an almost periodic solution, which always exists.

Theorem 6. *Suppose that $m=3$ and that there is a bounded convex set $\Omega \subset R^3$ such that for sufficiently large $t \geqslant 0$*

$$U(t, 0)\Omega \subset \Omega, \tag{14}$$

where $U(t, s)$ is the solving operator. Then (1) *has at least one almost periodic solution with a module of Fourier exponents in $\mathfrak{M}_{rat}$.*

The additional condition (14) holds automatically when equation (1) is dissipative: in this case the set P is compact, and for Ω we can take any sufficiently large ball.

In the proofs of Theorems 5 and 6 we shall use certain results about the structure of homogeneous spaces of small dimension; we first assemble them separately.

2. Let $X \subset R^3$ be a homogeneous space with respect to a compact connected transitive transformation group Γ. It is well known that X is one of the following homogeneous spaces: a point, a circle, a solenoid (that is, a one-dimensional locally unconnected set), the two-dimensional torus, the two-dimensional sphere.

Proposition 8. *If the group Γ is non-commutative, then X is a sphere.*

Proof. We must establish that for any homogeneous space other than a sphere Γ is commutative. For a proof of the result that to a torus (not necessarily two-dimensional) corresponds only a commutative group we refer the reader to the paper by Montgomery & Samelson [92]). Hence, we only need to eliminate the case of a one-dimensional space.

Let $\dim X = 1$. We fix an element $x_1 \in X$ and let $\mathscr{H}$ be a stationary subgroup. It is known that $\mathscr{H}$ is a Lie group (see Pontryagin [99],

Russian p. 349, English tr. p. 344) and that X is homeomorphic to the space $\Gamma/\mathcal{H}$ of left cosets. The group $\mathcal{H}$ is finite. Indeed, let $\mathcal{H}_0$ be the component of the unit element of $\mathcal{H}$. If $\mathcal{H}_0 \neq \{e\}$, then we can find an $x_0 \in X$ for which $X_0 = \mathcal{H}_0 x_0 \neq \{x_0\}$. Clearly, $X_0 \neq X$, and X is a connected homogeneous space. But a one-dimensional connected homogeneous space does not have proper connected homogeneous subspaces. From the fact that $\mathcal{H}$ is finite it follows that $\dim \Gamma = 1$; but then it easily follows that Γ is commutative. Indeed, if $uv = vu$, then by the Peter–Weyl theorem there exists a homomorphism $\tau: \Gamma \to T$, where T is a Lie group, such that $\tau(uv) = \tau(vu)$. Since $\dim \tau(\Gamma) \leqslant \dim \Gamma = 1$, $\tau(\Gamma)$ is one-dimensional and consequently it is a commutative Lie group. This contradiction proves Proposition 8 completely.

3. Proof of Theorem 6. By Proposition 7 every element $u \in E$ is generated by a continuous mapping $\phi: R^m \to P$. From (14) and the equality $T = T^+$ it follows that $\phi\Omega \subset \Omega$; in other words, u has at least one fixed point. We consider the collection Λ of all closed sets $L \subset R^3$ such that

(a) $u(L) = L$ for every $u \in G$ (G-invariance);

(b) every $u \in G$ has at least one fixed point belonging to L;

(c) if L contains a topological sphere, then it also contains its interior.

Since $P \cap \Omega \in \Lambda$ the set Λ is non-empty. We order Λ by inclusion, and by Zorn's lemma obtain a minimal element $\mathring{P} \in \Lambda$. Let $\mathring{G}$ be the restriction of G to the set $\mathring{P}$. The group $\mathring{G}$ is connected; we shall prove that it is commutative.

By assuming otherwise we obtain that G acts non-commutatively on some orbit $G(p_0) \subset \mathring{P}$. Since by Proposition 5 $G(p_0)$ is a sphere, in view of the minimality of $\mathring{P}$ and the property (c), the set $\mathring{P}$ is a three-dimensional ball A. Next, the set of all those $p \in \mathring{P}$ for which the group G_p is non-commutative is obviously open. Therefore the interior of the ball A contains a G-orbit which is a sphere, and we obtain a second closed ball $A_1 \subset A$. Since the boundary of this ball is G-invariant, then A_1 itself is G-invariant (this is easily seen by considering the transformations $u \in G$ as the limits of solving operators), which contradicts the minimality of $\mathring{P}$. Thus, the group $\mathring{G}$ is commutative and connected. In $\mathring{G}$ (just as in every connected compact commutative group) we can find an element g such that

$\overline{\{g^n\}} = \mathring{G}$ $(n = 0, \pm 1, \ldots)$.[5] By condition (*b*) the transformation g has a fixed point p_0, but then every transformation $u \in G$ leaves p_0 fixed. this proves Theorem 6.

4. Proof of Theorem 5. For $m = 2$ the only G-orbits are a circle and a point. Let Q denote the set of all G-stationary points. We consider the collection Λ of all compact sets $L \subset P$ such that

(*a*) $u(L) = L$ for any $u \in G$,

(*b*) if L contains a circle, then it also contains its interior.

If $Q \neq P$, then Λ is non-empty since the closed disc corresponding to a non-trivial G-orbit is an element of Λ. Consider a 'minimal' element $\mathring{p} \in A$. It follows immediately from minimality that $\mathring{p}$ consists of G-stationary points. Thus, we have proved that $Q \neq \emptyset$. We shall show that

$$\text{either } Q = \{p_0\} \text{ or } Q = P. \tag{15}$$

In the first case the module of the almost periodic solution $p_0(t)$ belongs to $\mathfrak{M}$ because the corresponding minimal fibre or E-orbit consists of a single point. In the second case every bounded solution is almost periodic and has a module in $\mathfrak{M}_{\text{rat}}$. Suppose that there is a non-trivial orbit $G(p_0)$. We consider an arbitrary point $p \in P$ lying outside this orbit. Since the orbits $G(p)$ depend continuously on p (in the Hausdorff metric) and since the set P is linearly connected (as the continuous image of the plane R^2), to p corresponds a G-orbit enveloping the orbit $G(p_0)$. Hence it follows that there exists a unique stationary point and the G-orbits of the other points are circles surrounding it. This proves (15).

It can happen that to a non-trivial G-orbit there corresponds an almost periodic solution $p_1(t)$; the module of such a solution does not belong to $\mathfrak{M}_{\text{rat}}$. Let $\mathfrak{M}_1$ denote the smallest module containing the modules $\mathfrak{M}(F)$ and $\mathfrak{M}(p_1(t))$. We shall prove that every bounded solution is almost periodic and its module belongs to $\mathfrak{M}_1$. With this aim, in (1) we make the substitution $p(t) = p_1(t) + q(t)$ and obtain $\dot{q} = F_1(q, t) = F(q + p_1(t), t) - F(p_1(t), t)$. This equation has two almost periodic solutions with modules belonging to $\mathfrak{M}_1$ (namely,

[5] *Proof.* Let $a(t)$ be an everywhere dense 1-parameter subgroup, and x_m the characters of the group $\mathring{G}$, $x_m(a(t)) = \exp(i\lambda_m t)$. We choose a $t_0 \in J$ such that $\exp(i\lambda_m t_0) \neq 1$ for every m. Then we can take $g = a(t_0)$. Indeed, if $G_1 = \{g^n\} \neq \mathring{G}$, then we can find a non-trivial character of $\mathring{G}$ that is trivial on G. This is excluded by the choice of t_0.

$q \equiv 0$ and $q = (p_0(t) - p_1(t))$, and consequently it is enough to refer to a property of the form (15).

9 Realisation of the principle of the stationary point under monotonicity conditions

Suppose that the condition for V-monotonicity holds and that equation (1) has a solution bounded on $t \geqslant 0$. First we assume that V has the form $V(p, q) = W(p - q)$, where W is a homogeneous convex function on R^m, or more briefly, W is a norm on R^m. But even under these conditions the existence of at least one almost periodic solution is unknown in the case $m \geqslant 5$. To ensure the existence of at least one almost periodic solution, independently of the dimension we impose the additional condition that the norm W is *strictly convex*.

Theorem 7. *If the W-norm is strictly convex, then equation* (1) *has at least one almost periodic solution with a module in $\mathfrak{M}$. When $m \leqslant 4$, the strict convexity condition is not required; here the module of an almost periodic solution belongs to $\mathfrak{M}_{\text{rat}}$.*

Proof. By Corollary 6, $\|p_1(t) - p_2(t)\|_W = \|p_1(0) - p_2(0)\|_W$ holds identically for any $p_1(0)$, $p_2(0) \in P$; hence we obtain that E is a group of W-isometric operators. Every element of the group G is generated by some mapping $\phi: R^m \to P$ (Proposition 7). From the monotonicity condition and the equality $T = T^+$ it follows that ϕ is a W-non-expansion operator. We denote the set of all 'generating' mappings by $\tilde{G}$. In particular, the unit element in G is generated by the retract $\omega: R^m \to P$ with $\omega(p) = p$ for $p \in P$.

1. We prove the first assertion in Theorem 5. We shall show that the set P is convex and the transformations $u \in E$ are affine on P. Let $p_1, p_2 \in P$, $p = (p_1 + p_2)/2$ and $q = \omega(p)$. Since

$$\|q - p_1\|_W \leqslant \|p - p_1\|_W = \tfrac{1}{2}\|p_1 - p_2\|_W$$

and

$$\|q - p\|_W \leqslant \|p - p_2\|_W = \tfrac{1}{2}\|p_1 - p_2\|_W,$$

it follows immediately from the strict convexity of the W-norm that $q = \omega(p) = p$. In other words, the set P is convex. Similarly, it can be established that the transformations $u \in E$ are affine on P. Now we shall prove that the transformations $u \in E$ have a common fixed point p_0. It will then follow that the minimal fibre $E(p_0)$ consists of a single point, and so our problem reduces to Favard's theorem.

Let μ be an invariant measure on the compact group E. We consider a mapping $\phi: E \to R^m$, where $\phi(u) = u(p_1)$ (p_1 is a fixed point in E). It is easy to see that $p_0 = \int_E \phi(u)\, d\mu$ is an E-stationary point.

2. Now we prove the theorem for the case $m \leq 4$.

We assume that a W-norm is defined on R^m; all geometrical notions (distance, diameter, isometry) relate to the W-norm. We denote by $S(p, r)$ the W-ball with centre at the point p and radius r. We consider the set Λ of all convex compact sets $A \subset R^m$ such that $\tilde{G}(A) \subset A$, and order Λ by inclusion. The set Λ is non-empty. Indeed, let L be an arbitrary compact G-invariant set, for instance, $L = G(p)$, and set $\tilde{L} = \bigcap_{p \in L} S(p, d)$, where $d = \operatorname{diam} L$. Then it is not difficult to see that $\tilde{L}$ is convex, $L \subset \tilde{L}$, $\operatorname{diam} L = \operatorname{diam} \tilde{L}$, and $\tilde{G}(\tilde{L}) \subset \tilde{L}$. Let $\mathring{A}$ be a 'minimal' element of the set Λ. We assume that $\dim \mathring{A} = 4$, since if $\dim \mathring{A} \leq 3$ it is not difficult to repeat the arguments of Theorem 3. We study the set $\mathring{A}$ in greater detail.

Let $\partial \mathring{A}$ denote the boundary of $\mathring{A}$. We set $d_0 = \operatorname{diam} \mathring{A}$ and $\mathring{P} = \mathring{A} \cap P$. The inclusion $\mathring{P} \subset \partial \mathring{A}$ is an important consequence of 'minimality'. To prove this we take an arbitrary point $p_0 \in \mathring{P}$ and note that it follows directly from 'minimality' that $\operatorname{diam} G(p_0) = d_0$. Therefore, we can find points $p_1, q_1 \in G(p_0)$ for which $\|p_1 - q_1\|_W = d_0$. Since the group G acts transitively and isometrically on $G(p_0)$ we can find a $q_0 \in \partial \mathring{A}$ such that $\|q_0 - p_0\|_W = d_0 = \operatorname{diam} \mathring{A}$, that is, $p_0 \in \partial \mathring{A}$, which proves the inclusion $\mathring{P} \subset \partial \mathring{A}$.

Two cases are logically possible: (1) $\mathring{P} = \partial \mathring{A}$, (2) $\mathring{P} \neq \partial \mathring{A}$. The first of these cannot be realised since it would mean the existence of a retract $\omega: \mathring{A} \to \partial \mathring{A}$. When $\mathring{P} \neq \partial \mathring{A}$ we reason in exactly the same way as in Theorem 6, with the set $\mathring{A}$ playing the role of Ω. The connected orbits have the same structure since $\mathring{P}$ belongs to a part of $\partial \mathring{A}$ homeomorphic to a three-dimensional ball. This proves Theorem 7.

Theorem 7 can be given a more general form without essential changes in the proof. We call $V(p, q)$ (strictly) convex if it is (strictly) convex with respect to the first variable for a fixed value of the second. Then for $m \leq 4$ the condition of V-monotonicity with a convex V is sufficient, while for $m > 4$ V is strictly convex.

Comments and references to the literature

The monograph of Nemytskii & Stepanov [95], Chapter 5, contains a very full account of the general theory of dynamical systems (including 'topological dynamics'). The book by Gottschalk & Hedlund [38] is also interesting.

§ 1. The concept of a distal system is due to Ellis [119] who evidently used the important work on flows on a multi-dimensional torus by Furstenberg [108]. An approach to a non-autonomous equation from the point of view of dynamical systems was given by Millionshchikov [88] and also by Shcherbakov [114]. Later (and apparently independently) similar ideas were developed by American mathematicians (Miller, Sell, and others).

§ 2. The separation lemma and also the majority of its applications is given in the article by Zhikov [50]. It must be mentioned that the method of proving this lemma is a generalisation of a method of Ellis [119]. Proposition 2 was proved by Wallace [105]. Proposition 4 (in the case when all the fibres are distal) was proved by Furstenberg [109] and then generalised by Bronshtein [31].

§§ 3 and 4. The criterion for absolute recurrence is a generalisation of a similar criterion of Flor [110]. The remaining material in §§ 4 and 5 is due to Zhikov. We mention that Bronshtein & Chernyi [32] have obtained certain further results.

§ 5. Example 2 is taken from the article of Zhikov [50].

§§ 7–9. The formulation and application of the principle of the stationary point were given by Zhikov [50]. It appears that a study of higher dimensions by topological methods is interesting. We also note that the concept of V-monotonicity in a general form has been studied by Cheresiz [113].

8 Favard theory

1 Introduction

In this chapter we consider the problem of almost periodic solutions of a general linear equation in a Banach space

$$Lu \equiv u' + A(t)u = f(t). \tag{1}$$

In the simplest case, when $A(t) \equiv 0$, we obtain the problem about the integral we studied in Chapter 6, which will serve as a natural motivation for the construction of a more general theory. The Banach space B is assumed to be separable and the operators $A(t)$ unbounded. We do not need more precise information concerning the nature of these operators since we shall rely only on the properties of a solving Cauchy operator. In other words, the conditions on the operator $A(t)$ (and also on the way it depends on t) are implicit in the conditions on solving operators.

We assume that $A(t)$ and $f(t)$ are almost periodic functions with values in certain metric spaces. This enables us (at least formally) to define the set $\mathcal{H}$ of all 'limiting' pairs $h = \{\hat{A}(s), \hat{f}(s)\}$ of the form $\hat{A}(s) = \lim_{m\to\infty} A(s+t_m)$, $\hat{f}(s) = \lim_{m\to\infty} f(s+t_m)$. We denote the initial pair $\{A(s), f(s)\}$ by h_0 and identify it with the original equation (1). For any $h = \{\hat{A}(s), \hat{f}(s)\}$ we introduce the corresponding 'limiting' equation

$$\hat{L}u \equiv u' + \hat{A}(t)u = \hat{f}(t). \tag{1_h}$$

We rely on the following condition.

Condition 1 (Condition of 'weak' continuity). *For any initial value $p \in B$ and any $h \in \mathcal{H}$, equation (1_h) has a unique strongly continuous* *solution* $u(t) = S_h(t)p$, *and the map*

$$S_h(t): B \times \mathscr{H} \to B$$

is continuous for each $t \geq 0$ *if the space* B *is equipped with the weak topology on each bounded set.*

In other words, Condition 1 means that for any equation (1_h) the Cauchy problem is uniquely right-solvable for positive t for any initial value $u(0) \in B$, and that this solution depends continuously both on the initial condition and on the variation of the 'coefficients' $\hat{A}(t)$, $\hat{f}(t)$ at the limits of the set $\mathscr{H}$.

We define a transformation $S(t): X \to X (t \geq 0)$ on the space $X = B \times \mathscr{H}$ by the formula $S(t)x = S(t)\{p, h\} = \{S_h(t), h^t\}$. It is not difficult to verify that the operators $S(t)$ $(t \geq 0)$ commute. In what follows the space B will be endowed with the weak topology on bounded sets. The corresponding topology on $X = B \times \mathscr{H}$ will, for simplicity, also be called *weak*. From Condition 1 we obtain that $S(t)$ $(t \geq 0)$ is a semigroup of continuous transformations. It is also clear that if we give Condition 1 a two-sided character (by allowing $t < 0$), then the transformations $S(t)$ give a group of homeomorphisms. But two-sided conditions are very restrictive, and we shall not assume this. We also need the following condition.

Condition 2. *Equation* (1) *has at least one solution weakly compact on* $t \geq 0$.

If $u(t)$ $(t \geq 0)$ is a weakly compact solution in Condition 2, then the corresponding trajectory $\{u(t), h_0{}^t\}$ is weakly compact. Consequently (for instance, by Birkhoff's theorem (see Chapter 1, § 4)), there exists a weakly compact invariant subset $\mathring{X} \subset B \times \mathscr{H}$. Since the space B is separable and $\mathscr{H}$ is a compact metric set, the topology of convergence on X is metrisable (see Dunford & Schwartz [40], p. 426). Therefore, we can apply Proposition 2 of Chapter 1; as a result we obtain

Proposition 1. *Let* $\mathring{X} \subset B \times \mathscr{H}$ *be a weakly compact invariant subset and* $\{u_h(t), h^t\}$ *an arbitrary trajectory in* $\mathring{X}$. *Then the family of solutions* $\{u_h(t)\}$ *is weakly uniformly continuous on the entire axis.*

A trajectory in $\mathring{X}$ has the form $\{u_h(t), h^t\}$, where $u_h(t)$ is a solution of (1_h) defined on the whole time axis. If we fix $h \in \mathscr{H}$, then the set of corresponding initial values $\{u_h(0)\}$ is called a *fibre over h* and is denoted by $\mathring{X}_h$.

The problem of almost periodic solutions splits in an obvious way into two separate problems: the problem of weak almost periodicity and the problem of compactness. Each of these two problems in turn requires an individual approach, and has in a certain sense an independent character. In questions of weak almost periodicity we shall rely on the following 'simplest' principles.

Theorem 1. *If each fibre $\mathring{X}_h$ ($h \in \mathcal{H}$) consists of one element, then all the trajectories in $\mathring{X}$ are weakly almost periodic.*

Theorem 2. *If for any $t \in J$ the fibre $\mathring{X}_{h_0 t}$ consists of one element, then the corresponding trajectory is weakly N-almost periodic.*[1]

Theorems 1 and 2 are special cases of Theorems 1 and 2 in Chapter 7, we only need to replace compactness by weak compactness.

We end this subsection by mentioning a fairly simple case when Condition 1 holds.

Let the operator $A(t)$ have the form $A(t) = A_0 + A_1(t)$, where A_0 does not depend on t and the bounded operator $A_1(t)$ is almost periodic relative to the operator norm. Suppose that A_0 generates a strongly continuous semigroup $U(t) = \exp(-A_0 t)$ ($t \geqslant 0$), and that the free term f is an element of the space $\mathring{M}_1(B)$. In this case the space $\mathcal{H}$ is a collection of pairs $\{\hat{A}_1(s), \hat{f}(s)\}$ with the metric

$$d(\hat{h}, \hat{\hat{h}}) = \sup_{s \in J} \|\hat{A}_1(s) - \hat{\hat{A}}_1(s)\| + \sup_{s \in J} \int_0^1 \|\hat{f}(\tau + s) - \hat{\hat{f}}(\tau + s)\| \, d\tau.$$

We determine the solution of the non-homogeneous equation (1_h) from the following integral equation:

$$S_h(t)p = u(t) = U(t)p + \int_0^t U(t-s)\{\hat{A}_1(s)u(s) + \hat{f}(s)\} \, ds. \tag{2}$$

It can be established by the usual method of successive approximations that there exists a unique solution on $[0, T]$, and that it satisfies the estimate

$$\|S_h(t)p\| \leqslant c_1(T)\|p\| + c_2(T) \quad (0 \leqslant t \leqslant T),$$

[1] When the two-sided uniqueness theorem holds for (1) it is sufficient that the fibre $\mathring{X}_{h_0}$ consists of one element.

where the constants $c_1(T)$ and $c_2(T)$ do not depend on $h \in \mathcal{H}$. We prove that the mapping $S_h(t): B \times \mathcal{H} \to B (t \geqslant 0)$ is continuous in the appropriate sense. Since the operator $U(t)$ $(t \geqslant 0)$ is weakly continuous, it is sufficient to verify that the integral term in (2) depends continuously on $h \in \mathcal{H}$ for each $t \geqslant 0$. The latter is easily obtained from the above estimate and from the definition of the metric d.

2 Weak almost periodicity (the case of a uniformly convex space)

The problem of weak almost periodicity turns out to be closely connected with the geometric properties of the Banach space B. Here we show that in the case of a uniformly convex space, a weakly almost periodic solution can be found by a simple 'minimax' method. We assume that Conditions 1 and 2 hold, and denote weak convergence in B by the symbol $\rightharpoonup$.

We say that equation (1) satisfies the *condition of semiseparation* if the non-trivial bounded solutions of the homogeneous equation are semi-isolated away from zero, that is, if $\lim_{t\to-\infty} \|u(t)\| > 0$ whenever $L(u) = 0$, $u(t) \not\equiv 0$, $\sup_{t \in J} \|u(t)\| < \infty$. Note that non-trivial solutions of the homogeneous equation can vanish since we do not assume that the left uniqueness theorem holds; if we were to make this assumption, then semi-isolation away from zero would reduce to the inequality $\inf_{t \leqslant 0} \|u(t)\| > 0$.

Theorem 3. *Let the space B be uniformly convex. Then if the condition of semiseparation holds for the original equation* (1), *then it has at least one weakly N-almost periodic solution. This solution is weakly almost periodic if each equation* (1_h) *satisfies the condition of semiseparation.*

Proof. We introduce the quantity

$$\mu(h) = \inf \sup_{t \in J} \|u(t)\|,$$

where the inf is taken over all bounded solutions of (1_h). By taking a minimising sequence and extracting from it a subsequence weakly convergent for each $t \in J$,[2] we see that the inf is attained on some solution $u_h(t)$, that is,

$$\sup_{t \in J} \|u_h(t)\| = \mu(h). \tag{3}$$

[2] The possibility of taking the weak limit (uniform on each finite segment) is ensured by Proposition 1.

We are going to show that $\mu(h) \equiv \mu$. Assume that $\mu(h_1) < \mu(h_2)$. We find a sequence $\{t_m\} \subset J$ for which $h_1^{t_m} \to h_2$. By taking a subsequence if necessary, we can assume the weak convergence $u_{h_1}(t+t_m) \rightharpoonup v(t)$. But $v(t)$ is a solution of (1_{h_2}) and

$$\sup_{t\in J} \|v(t)\| \leqslant \sup_{t\in J} \|u_{h_1}(t+t_m)\|$$

$$\leqslant \sup_{\mu(h_2)} \|u_{h_1}(t+t_m)\| = \mu(h_1) < \mu(h_2),$$

which contradicts the definition of the number $\mu(h_2)$. Thus $\mu(h) \equiv \mu$.

Let $\mathring{X}$ denote the set of all trajectories of the form $\{u_h(t), h^t\}$, where h ranges over the set $\mathscr{H}$ and the solution $u_h(t)$ satisfies the minimax condition (3). Then, a similar argument shows that $\mathring{X}$ is weakly compact.

Now we begin to prove the first part of the theorem. We recall that a space B is called *uniformly convex* if the inequality

$$\|p_1 - p_2\| \geqslant \delta \max\{\|p_1\|, \|p_2\|\}$$

implies that

$$\tfrac{1}{2}\|p_1 + p_2\| \leqslant (1-\phi(\delta)) \max\{\|p_1\|, \|p_2\|\} \quad (\phi(\delta) > 0).$$

We shall prove that the unique solution of (1) satisfies the minimax condition (3). Indeed, suppose that $u_0(t)$ and $u_1(t)$ are two such solutions. If $u_0(t) \not\equiv u_1(t)$, then by the condition of semiseparation we can find a $t_0 \in J$ such that

$$\inf_{t\leqslant t_0} \|u_0(t) - u_1(t)\| = \delta\mu > 0.$$

We choose a subsequence $t_m \to -\infty$ for which $h_0^{t_m} \to h_0$. Without loss of generality we may assume that

$$u_0(t+t_m) \rightharpoonup v_0(t),$$

$$u_1(t+t_m) \rightharpoonup v_1(t) \quad (t \in J).$$

Then for any $t \in J$ we have

$$\tfrac{1}{2}\|v_0(t) + v_1(t)\| \leqslant \tfrac{1}{2} \lim_{m\to\infty} \|u_0(t+t_m) + u_1(t+t_m)\|$$

$$\leqslant (1-\phi(\delta))\mu < \mu,$$

which contradicts the definition of μ. Thus we obtain that the fibre $\mathring{X}_{h_0}$ (and also the fibre $\mathring{X}_{h_0 t}$) consists of one element. By Theorem 2, to this fibre there corresponds a weakly N-almost periodic solution.

For a proof of the second part of the theorem we need to prove similarly that each fibre $\mathring{X}_h$ consists of a single element, and to use

Theorem 1. Then, Theorem 3 is proved. It remains to verify that the solution given by Theorem 3 is, in fact, not necessarily compact. Let H be a separable Hilbert space. We denote by B the Hilbert space of measurable functions $u: J \to H$ with the norm

$$\|u\|_B^2 = \int_J \|u(s)\|^2 Q(s)\, \mathrm{d}s,$$

where the weight function $Q(s)$ is continuous and positive, and $Q(s) \sim 1/s$ as $s \to \infty$. Let C be a Hermitian completely continuous operator $H \to H$ with a complete system of eigenvectors $\{e_m\}$ corresponding to the positive eigenvalues $\{\lambda_m\}$.

For each $u(s) \in B$ the translation $u(t+s)$ is a continuous function $J \to B$; this function is, obviously, a 'generalised' solution of the equation $u'_t - u'_s = 0$. We now fix a weakly almost periodic but not compact function $\mathring{u}(s): J \to H$ (an example is given at the end of Chapter 4). It is easy to verify that to it there corresponds a weakly almost periodic but not compact function $\mathring{u}(t+s): J \to B$. The function $\mathring{u}(t+s)$ satisfies the equation $u'_t - u'_s + Cu = f(t+s)$, where $f(t) = C\mathring{u}(t)$; this equation is considered in the space B and has the form $u'_t + Au = g(t)$, where $A = -\mathrm{d}/\mathrm{d}s + C$ and $g(t) = f(t+s)$. Thus, the equation has a weakly almost periodic solution; we are going to show that the homogeneous equation $u'_t + Au = 0$ has no bounded solutions.

Indeed, let $u(t, s): J \to B$ be a non-trivial solution of the homogeneous equation. We set $u = \sum u_m e_m$ and assume that $u_m(t, s) \not\equiv 0$. Then

$$\frac{\partial}{\partial t} u_m - \frac{\partial}{\partial s} u_m + \lambda_m u_m = 0.$$

If now we set $v_m(t, s) = u_m(t, s) \exp(-\lambda_m t)$, then $v_m(t, s)$ satisfies the relation

$$\frac{\partial}{\partial t} v_m - \frac{\partial}{\partial s} v_m = 0,$$

and so v_m must have the form $v_m = v_m(t+s)$. On the other hand (since u is a bounded solution and $\lambda_m > 0$), the norm $\|v_m(t+s)\|_B^2 = \int \|v_m(t+s)\| Q(s)\, \mathrm{d}s$ decreases exponentially as $t \to +\infty$. But this is impossible (here we must use the condition $Q(s) \sim 1/s$). This, then, is the required example. Clearly, the operator A can be chosen to be bounded; for this we must replace B by a subspace corresponding to a fixed width of the spectrum.

3 Certain auxiliary questions

From now on, unless stated otherwise there will be no restrictions on the space B (apart from separability).

1. Construction of a minimal set. We consider the set of all weakly compact convex sets $K \subset B$ such that $K \times \mathcal{H}$ contains at least one trajectory of the basic semigroup $S(t)$,[3] and order this set by inclusion. Every minimal element of this set is called a *Favard minimal set.*

Let K be a Favard minimal set. The set of all trajectories belonging to $K \times \mathcal{H}$ is denoted by F, and the fibre over $h \in \mathcal{H}$ by F_h. It is obvious that the fibre F_h is a convex weakly compact set. Let ρ be the metric corresponding to the topology of weak convergence on K. We shall prove that for any elements $p(0), q(0) \in F_h$

$$\inf_{t \leqslant 0} \rho(p(t), q(t)) = \inf_{t \in J} \rho(p(t), q(t)) = 0. \tag{4}$$

With this aim we embed our space B in a Hilbert space H so that every weakly convergent sequence in B is strongly convergent in H.[4] Then, in particular, the weak topology on K is equivalent to the strong topology on H. Now we show that

$$\sup_{t \in J} \|u(t)\|_H = \sup_{t \leqslant 0} \|u(t)\|_H \equiv \mu \tag{5}$$

for any trajectory $\{u(t), h^t\} \subset F$. Suppose, for instance, that there are trajectories $x_1(t) = \{u_1(t), h_1{}^t\}$ and $x_2(t) = \{u_2(t), h_2{}^t\}$ for which $\alpha = \sup_{t \leqslant 0} \|u_1(t)\|_H < \sup_{t \in J} \|u_2(t)\|_H$. We consider the set $K_1 = \{p \in K : \|p\|_H \leqslant \alpha\}$. Then the set $K_1 \times \mathcal{H}$ contains the semitrajectory $\{x_1(t)\}$ $(t \leqslant 0)$ but does not coincide with $K \times \mathcal{H}$, which is impossible. Now the property (4) follows from (5). For suppose, for instance, that $\inf_{t \leqslant 0} \|p_1(t) - p_2(t)\|_H > 0$. Then for the half-sum $(p_1(t) + p_2(t))/2$ we have (by the parallelogram equality in H) $\sup_{t \leqslant 0} \|(p_1(t) + p_2(t))/2\|_H < \mu$, which contradicts (5).

Proposition 2. *Suppose that equation* (1) *has a (weakly) compact solution on* $t \geqslant 0$. *Suppose that the non-trivial (weakly) compact solutions of the homogeneous equations* $\hat{L}u = 0$ *are (weakly) semi-*

[3] For this it is sufficient that the set $K \times \mathcal{H}$ contains at least one semitrajectory.

[4] For this it is enough to realise B as a subspace of $C(0, 1)$ and take H to be $\mathcal{L}^2(0, 1)$.

separated from zero. Then (1) *has at least one* (*weakly*) *almost periodic solution.*

Proof. *In the 'weak' case each fibre* F_h is trivial, as follows at once from (4). In the 'strong' case we must notice that a compact Favard minimal set exists, and then property (4) holds in the sense of the norm. The matter is then reduced to Theorem 1. It also follows from property (4) that the set F contains a unique Birkhoff minimal subset; we denote it by $\mathring{F}$.

2. Some properties of extreme points. It is known that if P is a weakly compact set in B, then every extreme point of its closed convex hull belongs to P and so is an extreme point of P (see Dunford & Schwartz, [40], p. 440).

Let K be a Favard minimal set, and i denote the natural projection $B \times \mathcal{H} \to B$.

We call a point $x = \{p, h\} \in F$ *extreme* if the component p is an extreme point of the fibre F_h. From the right uniqueness theorem it follows that the point $x(t) = \{p(t), h^t\}$ is also extreme for $t \geqslant 0$. Hence it follows that the set of extreme points form an everywhere dense subset of $\mathring{F}$. Indeed, if we assume otherwise, then (by taking into account that in a minimal set every semitrajectory is everywhere dense) $\mathring{F}$ does not contain any extreme points at all. But it follows from Favard minimality that K is the convex closure of each of the following weakly compact sets

$$i(F) = \bigcup_{h \in \mathcal{H}} F_h, \quad i(\mathring{F}) = \bigcup_{h \in \mathcal{H}} \mathring{F}_h.$$

Hence it follows that every extreme point of K is an extreme point of some fibre F_h and belongs to its subset $\mathring{F}_h$. This contradiction proves our assertion.

The next subsection is devoted to a general study of the important property (4).

3. On completely non-distal extensions. Let X be a metric compact space on which a semigroup $S(t)$ ($t \geqslant 0$) acts invariantly. Let M be a compact set that is minimal with respect to the action of a dynamical system on it. We assume that the semigroup $S(t)$ is an extension of the dynamical system on M, that is, there is a continuous mapping $j: X \to M$ such that

$$j(s(t)x) = h^t \quad (h = j(x), x \in X, t \geqslant 0).$$

We say that $S(t)$ is a *completely non-distal extension* if

$$\inf_{t \in J} \rho(x_1(t), x_2(t)) = 0 \tag{6}$$

whenever $j(x_1(0)) = j(x_2(0))$.

The main problem regarding completely non-distal extensions is the following: is there at least one trivial (that is, one element) fibre X_h? Unfortunately, the solution to this problem in such a general form seems to be negative. Here we give an affirmative solution of the problem under a certain additional condition.

We call a fibre X_{h_0} *positively stable* if the transformations $S(t)$ $(t \geqslant 0)$ are equicontinuous on it.

Lemma 1. *Let X_{h_0} be a positively stable fibre. Then the trivial fibres form a set of the second category in M.*

Proof. For convenience of notation we denote the fibre $X_h (h \in M)$ by $\Gamma(h)$. We consider the function $g(h) = \text{diam}\,(\Gamma(h))$ (the diameter of $\Gamma(h)$). Since this function is semicontinuous on M it has a non-empty set Ω of points of continuity (Ω is a set of the second Baire category). We show that $g(h) = 0$ for $h \in \Omega$. By assuming otherwise we obtain $g(h) \geqslant k_0 > 0$ for $h \in \Lambda$, where Λ is an open set in M. Since M is minimal there is a finite set $\{t_i\} \subset J^+$ such that $M = \bigcup \Lambda_{t_i}$. From this we conclude easily that

$$g(h) \geqslant k_1 > 0 \quad (h \in M). \tag{7}$$

Let $x_1, x_2, \ldots, x_m$ be an arbitrary finite collection of elements in the single fibre $\Gamma(h)$. Then

$$\inf_{t \geqslant 0} \text{diam}\,\{x_i(t)\} = 0. \tag{8}$$

For the proof let Z denote the product of m copies of X. Then if (8) does not hold, the closure of the semitrajectory $\{x_1(t), \ldots, x_m(t)\}$ $(t \geqslant 0)$ is separated from the diagonal in Z. But then the minimal subset contained in this closure is also separated from the diagonal. As is easily seen, this means that the original set X contains at least two different minimal sets, which is impossible in view of property (6). It follows from the positive stability of a fibre $\Gamma(h_0)$ that for any $\varepsilon > 0$ we can find a finite set $\{x_i(0)\} \subset \Gamma(h_0)$ such that $\{x_i(t)\}$ is an ε-net of $\Gamma(h_0{}^t)$ for any $t \geqslant 0$. But then from (8) we find that

$$\inf_{t \geqslant 0} \text{diam}\,\Gamma(h_0{}^t) \leqslant \varepsilon,$$

which contradicts (7). This proves Lemma 1.

4. The finite-dimensional case. Let the space B be finite-dimensional.

Corollary 1. *There is a set $\Omega \subset \mathcal{H}$ of the second Baire category such that for $h \in \Omega$ equation (1_h) has at least one N-almost periodic solution.*

Proof. We consider the set F. If we can prove that every fibre is positively stable (it is sufficient to do this for one fibre), then the problem is reduced to Lemma 1 and Theorem 2. But the stability of each fibre is an immediate consequence of finite-dimensionality. Indeed, consider the hyperplane Q defined by

$$Q = \{u(0) \in B : Lu = f, \sup_{t \in J} \|u(t)\| < \infty\}.$$

The transformations $S_{h_0}(t)$ $(t \geqslant 0)$ are uniformly bounded on Q, and being affine, they are equicontinuous.

In §7 we give an example which shows that the set Ω does not in general coincide with $\mathcal{H}$ (this example is important in a number of other connections). We also remark that the question of the validity of Corollary 1, for instance, for a Hilbert space, is open.

5. On an equivalent norm. It is known (see Kadets [62]) that a separable Banach space B can be given an equivalent norm with the property: if $x_m \to x$ and $\|x_m\| \to \|x\|$, then $\|x - x_m\| \to 0$.

Lemma 2. *Let K be a metric compact set and a function $\phi : K \to B$ be weakly continuous. Then the points of strong continuity form a set of the second category in K.*

Proof. We assume that an equivalent norm with the above property has been introduced on B. Now $\|\phi\|$ is lower semicontinuous, and so every point of continuity of it is, obviously, a point of strong continuity of ϕ, as we required.

6. Remarks on semigroups. Let $\mathring{X}$ be a compact space with a semigroup $S(t)$ acting invariantly and minimally. We need the following elementary fact: if Δ is an open set in $\mathring{X}$, then we can find a finite set $\{t_i\} \in J^+$ for which $\mathring{X} = \bigcup S(t_i)\Delta$. For the proof we set $t\Delta = S(t)\Delta$ for $t \geqslant 0$ and $t\Delta = \{x \in \mathring{X} : S(-t)x \in \Delta\}$ for $t < 0$. Obviously, the sets $t\Delta$ are open for $t \leqslant 0$. From the condition of minimality we obtain easily that $\mathring{X} = \bigcup t\Delta$; by choosing a finite covering we have $\bigcup t_i\Delta = \mathring{X}$. It follows from minimality that $S(t)\mathring{X} = \mathring{X}(t \geqslant 0)$. Therefore $\mathring{X} =$

$\bigcup(t_0+t_i)\Delta$ for any $t_0\geqslant 0$. By taking t_0 sufficiently large we obtain the required property.

4 Weak almost periodicity (the general case)

Theorem 3 was proved by a geometrical method based on uniform convexity. Nevertheless, a similar assertion can be proved even in the general case.

Theorem 4. *If the condition of semiseparation holds for equation (1), then the Favard minimal set contains a weakly N-almost periodic solution. This solution is weakly almost periodic if the condition of semiseparation holds for each equation* (1_h).

Proof. We consider the minimal set $\mathring{F}$ introduced in § 3.1 (subsection 1 of § 3), denoting the fibre over $h\in\mathscr{H}$ by $\Gamma(h)$. We prove that the fibres $\Gamma(h_0{}^t)$ $(t\in J)$ are trivial. If this were not the case, then we would have two corresponding solutions $p_1(t)\not\equiv p_2(t)$ of (1). By the condition of semiseparation we have

$$\inf_{t\leqslant t_0}\|p_1(t)-p_2(t)\|=\delta>0. \tag{9}$$

We regard the projection $i:\mathring{F}\to B$ as a weakly continuous mapping of the compact set $\mathring{F}$ into B. By Lemma 2, there is at least one point of strong continuity. This means that for any positive $\varepsilon\leqslant\delta$ there is an open neighbourhood $\Delta=\Delta(\varepsilon)$ such that

$$\|i(x_1)-i(x_2)\|\leqslant\varepsilon/2 \quad (x_1, x_2\in\Delta). \tag{10}$$

We consider the solution $p(t)=(p_1(t)+p_2(t))/2$ and let X be the closure of the semitrajectory $\{p(t), h_0{}^t\}$ $(t\geqslant 0)$ in F. As was mentioned in § 3.2, the extreme points form an everywhere dense set in $\mathring{F}$. We choose an extreme point $x=\{q(0), h\}$ belonging to the open set Δ and show that $x\notin X$. By assuming otherwise we obtain

$$p(t_m)=\tfrac{1}{2}(p_1(t_m)+p_2(t_m)) \rightharpoonup q(0), \quad h_0{}^{t_m}\to h$$

for some sequence $t_m\to-\infty$. Without loss of generality we may assume that $p_1(t_m)\rightharpoonup q_1$ and $p_2(t_m)\rightharpoonup q_2$ weakly. Since $q(0)$ is an extreme point of the fibre F_h, we have $q_1=q_2=q$. But then, for large m, the points $\{p_1(t_m), h_0{}^{t_m}\}$ and $\{p_2(t_m), h_0{}^{t_m}\}$ belong to Δ, which is incompatible with (9) and (10). Thus, $x\notin X$. But then the Birkhoff minimal set in X also does not contain x, and consequently, does not coincide with $\mathring{F}$. This (as we noted in § 3.1) contradicts property (4), and so we have proved that the fibres $\Gamma(h_0{}^t)$ are trivial. It remains to apply Theorem 2. The second part of Theorem 4 is proved similarly. This completes the proof.

Now we make several remarks.

1. At the expense of some complication of the preceding proof we can establish that not only $\Gamma(h_0)$ but also F_{h_0} is trivial. This means that the Favard minimal set does not contain any solutions of (1) apart from the unique weakly N-almost periodic solution. A similar result in the minimax method is: every ball of minimal radius containing at least one solution of (1), actually contains no solutions other than the unique weakly N-almost periodic solution.

2. The condition of semiseparation can be stated in the following generalised form: if $u(t)$ is a bounded solution of the homogeneous equation (1), then we can find a sequence of intervals $T_n \subset J$ whose lengths increase to infinity, such that $\inf_{t\in T_n} \|u(t)\| > 0$.

3. We say that a bounded solution of equation (4) is *uniformly positively stable* if there is a constant $l > 0$ such that

$$\|u(t)\| \leq l\|u(t_0)\| \quad (t \geq t_0) \tag{11}$$

whenever $Lu = 0$, $u \in C$. It is clear that in this case the condition of semiseparation holds.

4. If the operator-function $A(t)$ does not depend on t or is periodic in t, then for the generalised condition of semiseparation to hold it is sufficient that one of the following manifolds is closed:

$$N = \{u(0) \in B : Lu = 0,\, u \in C\},$$
$$N^+ = \{u(0) \in B : Lu = 0,\, u \in C^+\},$$
$$N^- = \{u(0) \in B : Lu = 0,\, u \in C^-\}.$$

The proof follows directly from the Banach–Steinhaus theorem.

5 Problems of compactness and almost periodicity

The problem of compactness consists in finding a compact solution of (1) if a weakly compact solution is known to exist. In a narrower formulation the problem could be, for instance, the compactness of the weakly almost periodic solutions in Theorems 3 and 4. The compactness problem is non-trivial even for the simplest equation $u' = f(t)$ (this is clear from the discussion in Chapter 6).

To clarify our approach to problems of compactness, we consider what is in a certain sense the 'simplest' situation. Let $T(t)$ $(t \geq 0)$ be a strongly continuous semigroup of linear operators $B \to B$. Since linear operators are automatically weakly continuous, it is natural to endow B with the weak topology. Suppose that the semigroup $T(t)$ is uniformly bounded (that is, $\sup_{t\geq 0} \|T(t)\| < \infty$) and that $X \subset B$ is a weakly compact minimal[5] subset. We claim that the set X is compact in the strong sense, and consequently, by Markov's theorem it consists of almost periodic trajectories. The idea of the proof is as follows: according to Lemma 2 we must find a point of strong continuity of the identity mapping $X \to B$, and then use the minimality and uniform boundedness to 'carry over' the strong continuity to each point of X.

The approach to the compactness problem developed below is not based on the linearity of the equations but only on the weak continuity of the solving operators $S_h(t)$ $(t \geq 0, h \in \mathcal{H})$, which follows from linearity. Consequently, it makes sense to give the main results in a form that relates to general dynamical systems.

1. General lemmas on compactness. Let M be a compact metric space with a minimal dynamical system. We consider the weak topology on B; the corresponding topology on $B \times M$ is also called weak. Let a semigroup of continuous transformations $S_h(t)$ $(t \geq 0)$ be given on $B \times M$ that is an extension of the dynamical system on M, that is, having the form $S(t)\{p, h\} = \{S_h(t)p, h^t\}$. For $x = \{p, h\}$ we set $i(x) = p_1$, $j(x) = h$. Let d denote the metric on M.

We need a stronger form of the continuity condition.

Condition 3. *For any positive* ε, t_0 *and* K *there is a* $\delta = \delta(\varepsilon, t_0, K)$ *such that*

$$\|S_{h_1}(t_0)p - S_{h_2}(t_0)p\| \leq \varepsilon$$

whenever $d(h_1, h_2) \leq \varepsilon$ *and* $\|p\| \leq K$.

Lemma 3. *Let* $\Phi = \Phi(p, q)$ *be a metric on* B *that is strongly uniformly continuous on each bounded set. Suppose that the following conditions hold:*

(1) *the operators* $S_h(t)$ $(t \geq 0, h \in M)$ *satisfy Condition* 3;

(2) *the operators* $S_h(t)$ $(t \geq 0, h \in M)$ *are* Φ*-equicontinuous on each bounded set;*

(3) *on a weakly compact minimal set* $\mathring{X} \subset B \times M$ *the weak topology is subordinate to the topology defined by the metric* $\rho(x, y) = \Phi(i(x), i(y)) + d(j(x), j(y))$.

[5] Here and in what follows minimality is understood in the Birkhoff sense.

Then the weak topology on $\mathring{X}$ is equivalent to that defined by ρ.

By considering the special case of Lemma 3 when $\Phi(p, q) = \|p - q\|$ we obtain the following assertion.

Lemma 4 (Compactness lemma). *Suppose that the following conditions hold:*

(1) *there exists at least one weakly compact trajectory;*

(2) *the transformations $S_h(t)$ $(t \geqslant 0, h \in M)$ satisfy Condition 3:*

(3) *the transformations $S_h(t)$ $(t \geqslant 0, h \in M)$ are strongly equicontinuous on bounded sets.*

Then there exists a compact trajectory; moreover, every weakly recurrent trajectory is compact.

We begin by proving Lemma 4, which is simpler in form. Since the semigroup $S(t)$ has a weakly compact trajectory, there is a minimal weakly compact set $\mathring{X} \subset B \times M$. We denote the fibre over $h \in M$ by $\Gamma(h)$. Let $K = \sup_{x \in \mathring{X}} \|i(x)\|$. By using the fact that the operators $S_h(t)$ $(t \geqslant 0, h \in M)$ are strongly equicontinuous on sets bounded in B, we can choose an $\alpha = \alpha(\varepsilon)$ so small that

$$\sup_{t \geqslant 0, h \in M} \|S_h(t)p_1 - S_h(t)p_2\| \leqslant \varepsilon/2, \tag{12}$$

whenever $\|p_1 - p_2\| \leqslant \alpha(\varepsilon)$, $\|p_1\|, \|p_2\| \leqslant K$. By reasoning as in the proof of Theorem 4 we find an open set $\Delta = \Delta(\varepsilon)$ such that (10) holds. As we mentioned in § 3.4, there is a finite set $L = \{t_k\} \subset J^+$ such that $\mathring{X} = \bigcup_{t \in L} S(t)\Delta$. We put $\Delta_k = S(t_k)$. Then we use Condition 3 and choose a $\delta = \delta(\varepsilon)$ small enough that

$$\sup_{t \in L} \|S_{h_1}(t)p - S_{h_2}(t)p\| \leqslant \varepsilon/2 \tag{13}$$

whenever $d(h_1, h_2) \leqslant \delta(\varepsilon)$, $p \in i(\mathring{X})$. Since a translation on M is continuous, we can find a $\beta = \beta(\varepsilon)$ such that $d(h_1^{\,t}, h_2^{\,t}) \leqslant \delta(\varepsilon)$ if $t \in L$ and $d(h_1, h_2) \leqslant \beta$. Let $M = \bigcup M_s$ be a decomposition of M into a finite number of sets M_s with diameters $\operatorname{diam}(M_s) \leqslant \beta$. Then $\mathring{X} = \bigcup \mathring{X}_s$, where $\mathring{X}_s = \bigcup_{h \in M_s} \Gamma(h)$. We set $G^s{}_k = \mathring{X}_s \cap \Delta_k$, and prove that for arbitrary elements $x_1, x_2 \in G^s{}_k$

$$\|i(x_1) - i(x_2)\| \leqslant \varepsilon. \tag{14}$$

Since the sets $G^s{}_k$ form a finite covering of the space $\mathring{X}$, (14) means that $i(\mathring{X})$ is strongly compact. If $x_1, x_2 \in G^s{}_k$, then there is a $\tau \in L$ such that $x_1 = S(\tau)y_1$ and $x_2 = S(\tau)y_2$, where $y_1, y_2 \in \Delta$.

Putting $p_1 = i(y_1)$, $p_2 = i(y_2)$, $h_1 = j(y_1)$, and $h_2 = j(y_2)$ we obtain $d(h_1, h_2) \leq \delta(\varepsilon)$ and $\|p_1 - p_2\| \leq \varepsilon/2$. Now, by (12) and (13) we have

$$\begin{aligned}\|i(x_1) - i(x_2)\| &= \|S_{h_1}(\tau)p_1 - S_{h_2}(\tau)p_2\| \\ &\leq \|S_{h_1}(\tau)p_1 - S_{h_1}(\tau)p_2\| + \|S_{h_1}(\tau)p_2 - S_{h_2}(\tau)p_2\| \\ &\leq \varepsilon/2 + \varepsilon/2 = \varepsilon.\end{aligned}$$

This proves Lemma 4.

Lemma 3 is proved in a completely similar way; we merely make a few observations. Obviously, Lemma 3 would be proved if we could show that the set $\mathring{X}$ endowed with the metric ρ is compact. By arguing as in Lemma 4, instead of the estimate (14) we have

$$\Phi(i(x_1), i(x_2)) \leq \varepsilon \quad (x_1, x_2 \in G^s{}_k).$$

The continuity of $\Phi(i(x), i(x))$ on $\mathring{X}$ does not follow immediately from here because, in general, the sets $G^s{}_k$ are not open.[6] However, it follows from this estimate that $\mathring{X}$ is compact in the metric ρ, that is, this metric is topologically equivalent to the weak metric on $\mathring{X}$.

2. Some consequences. Lemmas 3 and 4 have a whole range of applications.

Definition. We say that the *condition of uniform positive stability* holds if (11) holds for any solution $u(t)$ $(t \geq t_0)$ of the homogeneous equation $Lu = 0$.

The property of uniform positive stability means that the operators $S_{h_0{}^\tau}(t)$ $(t \geq 0, \tau \in J)$ are equicontinuous on sets bounded in B. Since the set $\{h_0{}^t\}$ is dense in $\mathcal{H}$ and the mapping $S_h(t): B \times \mathcal{H} \to B$ is continuous, the operators $S_h(t)$ $(t \geq 0, h \in \mathcal{H})$ are also equicontinuous. In particular, the inequality (11) holds for solutions of $\hat{L}u = 0$.

Theorem 5. *Suppose that Conditions* 1–3 *and the condition of uniform positive stability hold. Then equation* (1) *has at least one almost periodic solution and every weakly recurrent solution is compact.*

For the proof we must use Lemma 4 together with Theorem 4. We consider the special case $A(t) \equiv 0$.

Corollary 2. *If the indefinite integral of an almost periodic function is weakly compact, then it is almost periodic.*

[6] In general, the sets Δ_k are not open.

Now we discuss in more detail the case when the operator A does not depend on t. The condition of uniform positive stability means that A generates a uniformly bounded semigroup of operators, that is

$$\sup_{t\geq 0} \|\exp(-tA)\| < \infty.$$

Corollary 3. *Every weakly recurrent solution of the non-homogeneous equation* (1) *is almost periodic.*

Proof. From Lemma 4 it follows that a weakly recurrent solution is compact. On the other hand, by Theorem 5 there exists an almost periodic solution. By taking their difference we obtain a compact solution of the homogeneous equation; by Markov's theorem it is almost periodic, as we required.

Corollary 4 (criterion for a point spectrum). *An operator A has a point spectrum on the imaginary axis if and only if the homogeneous equation $Lu \equiv u' + Au = 0$ has at least one weakly compact solution weakly separated from zero.*

Proof. First we prove the sufficiency. By Birkhoff's theorem, the homogeneous equation has a non-trivial weakly recurrent solution, which by Corollary 3 is almost periodic. But then the Fourier coefficients are the eigenvectors of A. To prove the necessity we note that if $Ax = i\lambda_0 x$ $(\lambda_0 \in J)$, then the solution $u(t) = x \exp(-i\lambda_0 t)$ is weakly separated from zero. We consider some examples which use the compactness theorem in this section together with the method of harmonic analysis (see Chapter 6, §§ 4, 5).

Example 1. Suppose that the operator A generates a uniformly bounded semigroup and that the resolvent of A is completely continuous. We consider the perturbed equation

$$u' + Au + A_1 u = g(t),$$

where A_1 is a compact operator and $g \in \mathring{C}(B)$. We shall prove that every weakly compact solution is almost periodic. The operator $A + A_1$ also has a completely continuous resolvent; therefore the spectrum of $A + A_1$ is rarified. Then by Theorem 5 of Chapter 6, a weakly compact solution $u(t)$ is weakly almost periodic. Furthermore, the function $A_1 u(t)$ is compact and, consequently, almost periodic. We set $g_1(t) = g(t) - A_1 u(t)$; then we have $u' + Au = g_1$. By applying Theorem 5 to this equation we obtain that the solution $u(t)$ is compact.

Example 2. In the Hilbert space H we consider the hyperbolic equation

$$u''+Q^2u+Q_1u=f,$$

where Q is a positive definite operator with a completely continuous inverse, and Q_1 is completely continuous relative to Q (this means that Q_1Q^{-1} is completely continuous).

Let W denote the Hilbert space of pairs $z=\{x, y\}$ with the norm $\|z\|_W^2=\|Qx\|^2+\|y\|^2$. By setting $z=\{u, u'\}$ we obtain

$$z'+Az+A_1z=g,$$

where $Az=\{-y, Q^2x\}$, $A_1z=\{0, Q_1x\}$, and $g=\{0, f\}$. The operator A generates in W a group of unitary operators, and A_1 is completely continuous; thus, the matter is reduced to Example 1, that is, every solution bounded in the 'energy' norm of W is almost periodic in this norm.

6 Weakening of the stability conditions

Stability type conditions played a part in all the results in the previous section. None the less, there is a class of linear problems for which stability conditions do not hold, but assertions of the type of the point spectrum criterion are still valid. We begin by stating a condition that replaces stability.

Let $A(t)\equiv A$ and $T(t)=\exp(-tA)$ $(t\geqslant 0)$. Instead of requiring the operators $T(t)$ $(t\geqslant 0)$ to be uniformly bounded we need a more general condition to be fulfilled: the function $v(t)=T^*(t)v(0)$ is bounded on $t\geqslant 0$ for a set of initial values $D=\{v(0)\}$ everywhere dense in B^*. As before, we denote by (ξ, p) the value of a functional $\xi\in B^*$ at $p\in B$. Under these conditions the following assertion holds.

Theorem 6. *Every compact* (*weakly recurrent*) *solution of the equation* $Lu=f$ *is* (*weakly*) *almost periodic*; *in particular, the criterion for a point spectrum holds.*

Proof. Since B is separable, the set D contains a countable total subset $\{\xi_i\}$; here we can assume that $\|\xi_i\|=1$.

We set $\xi_i(t)=T^*(t)\xi_i$ $(t\geqslant 0)$ and define on $B\times B$ the metric

$$\Phi(p, q)=\sum_{i=1}^{\infty}\sup_{\tau\geqslant 0}\frac{|(\xi_i(\tau), p-q)|}{2^i}.$$

It is not difficult to see that on each weakly compact set in B, Φ is not weaker than the metric of weak convergence. It is important that the operators $S_h(t)$ $(t\geqslant 0, h\in\mathcal{H})$ are Φ-non-expanding. Indeed, the difference $u(t)=S_h(t)p-S_h(t)q$ is a solution of the homogeneous equation, and therefore for $t\geqslant 0$ we have

$$\begin{aligned}\Phi(S_h(t)p, S_h(t)q) &= \sum_{i=1}^{\infty} \sup_{\tau\geqslant 0} \frac{|(\xi_i(\tau), u(t))|}{2^i} \\ &= \sum_{i=1}^{\infty} \sup_{\tau\geqslant 0} \frac{|(\xi_i(\tau+t), p-q)|}{2^i} \\ &\leqslant \Phi(p, q).\end{aligned}$$

First we study the more difficult case of a weakly recurrent solution. Let $p(t)$ be such a solution. Instead of $p(t)$ we consider a solution $q(t)$ lying in the set $\mathring{F}$ (see § 3.1). By Lemma 3 the metric $\rho(x, y)=\Phi(i(x), i(y))+d(j(x), j(y))$ is equivalent to the weak metric on $\mathring{X}$. But then any solutions from the same fibre $\mathring{X}_h$ must be weakly semiseparated, as follows at once from (15). Hence (with due regard for the fundamental property (4)), we conclude that each fibre $\mathring{F}_h$ is trivial, that is, $q(t)$ is a weakly almost periodic function.

The difference $u(t)=p(t)-q(t)$ is a weakly recurrent[7] solution of the homogeneous equation. On the closure of the trajectory $\{u(t)\}$ in B the metric Φ is equivalent to the weak metric (again by Lemma 3). By Markov's theorem $u(t)$ is weakly almost periodic, and so Theorem 6 is proved for weakly recurrent solutions.

The case of compact solutions is simpler. We first clarify the character of a compact solution of the homogeneous equation. It can be seen immediately that the metric Φ defines strong convergence on the closure of a compact trajectory. Therefore, a compact trajectory is almost periodic. The non-homogeneous equation is settled by referring to Proposition 2, and the theorem is proved completely.

Example 3. We consider the equation

$$\begin{bmatrix} p_1' \\ \vdots \\ p_m' \end{bmatrix} + \begin{bmatrix} A_{11} & \dots & A_{1m} \\ \vdots & & \vdots \\ A_{m1} & \dots & A_{mm} \end{bmatrix} \begin{bmatrix} p_1 \\ \vdots \\ p_m \end{bmatrix} = 0,$$

where the A_{ij} are commuting bounded normal operators in a Hilbert space; let H denote the product of m copies of this space. By relying on Theorem 6 we show that every compact (weakly recurrent) solution is (weakly) almost periodic. The operators A_{ij} are functions of a self-adjoint operator C, which splits into cyclic components, and

[7] Here we have used the fact that an almost periodic function is simultaneously recurrent with any recurrent function, see Chapter 7, § 3.

so we may assume that C is a cyclic operator. We realise C in the form of the operator of multiplication by λ in the space $\mathcal{L}^2(-1, 1, d\mu)$. Let

$$A = (A_{ij}), \quad \exp(-tA) = B(\lambda, t), \quad u = (p_1, \ldots, p_m).$$

We distinguish two cases. The first is when the bounded solutions of the equation $u' + Au = 0$ are dense in H. In this case the norm $\|B(\lambda, t)\|$ is bounded in $t \in J$ for almost all $\lambda \in [-1, 1]$ (in the sense of the measure μ); then the norm $\|B^*(\lambda, t)\|$ also has this property. Hence, in turn, it is easy to deduce that the set of bounded solutions of $u' + A^*u = 0$ is dense in H. In other words, Theorem 6 applies.

Next suppose that the closure of the bounded solutions is $H_1 \neq H$; let H_2 denote the orthogonal complement of H_1, and P_1, P_2 be the corresponding projection operators. It is easy to see that P_1 and P_2 are defined by operators of multiplication by measurable matrices. Since $u = P_1 u = u_1$, we find that $u_1' + P_1 A u_1 = u_1' + A_1 u_1 = 0$ in H_1, and we can apply the preceding arguments to this equation.

7 On solvability in the Besicovitch class

In this section we extend the original problem and study the solvability of (1) not in the class of Bohr almost periodic functions but in the wider class of Besicovitch almost periodic functions.[8] It turns out that in this wider setting the problem can be solved fairly simply and without stability or separation type conditions. We also give a construction of a linear equation that is solvable in the class of bounded Besicovitch almost periodic functions but not in the Bohr class.

1. Let μ be a normalised invariant measure on $\mathcal{H}$. We introduce the Lebesgue space B_1 consisting of μ-measurable mappings $\phi : \mathcal{H} \to B$ with the norm $\|\phi\|_{B_1}{}^2 = \int_{\mathcal{H}} \|\phi(h)\|^2 \, d\mu$.

Definition. A measurable mapping $\phi : \mathcal{H} \to B$ is called an *invariant section* if $\phi(h^t) = S_h(t)\phi(h)$ for any $t \in [0, \infty) = J^+$ almost everywhere on $\mathcal{H}$.

Let us explain this definition. It is obvious that $\phi(h^t)$ and $S_h(t)\phi(h)$ are measurable on $\mathcal{H} \times J^+$ and are equal almost everywhere. By

[8] A measurable function $f: J \to B$ is said to be *Besicovitch almost periodic* if for any $\varepsilon > 0$ there is a trigonometric polynomial $P_\varepsilon(t)$ such that

$$\overline{\lim_{T\to\infty}} \frac{1}{2T} \int_{-T}^{T} \|f(t) - P_\varepsilon(t)\|^2 \, dt \leq \varepsilon.$$

Fubini's theorem there is a set $\Gamma \subset \mathscr{H}$ of full measure such that $\phi(h^t) = S_h(t)\phi(h)$ for almost all $t \in J^+$, and after a suitable adjustment, for all $t \in J^+$. We set $\Gamma_1 = \bigcup_{t \geq 0} \Gamma^t$ and $\Gamma_0 = \bigcup_{t \geq 0} \Gamma_1{}^t$. Then Γ_0 is an invariant set, it has full measure, and $\phi(h^t) = S_h(t)\phi(h)$ for every $h \in \Gamma_0$ and $t \in J$. Consequently, $\phi(h^t)$ is a solution of equation (1_h) defined on the entire real line.

Theorem 7. *Under Conditions* 1 *and* 2 *there is at least one (essentially) bounded invariant section.*

Proof. We consider the family of operators $\Pi(t): B_1 \to B_1$ $(t \geq 0)$ defined by

$$\Pi(t)\phi(h) = S_{h^{-t}}(t)\phi(h).$$

It is easy to see that these operators commute. What is important for us is that they have the following property: an element $\phi \in B_1$ is an invariant section if it is a common fixed point for the operators $\Pi(t)$ $(t \geq 0)$. We consider the set F introduced in § 3.1, and denote by Z the set of all measurable mappings $\phi: \mathscr{H} \to B$ for which $\phi(h) \in F_h$. The set Z is convex (since the fibres F_h are convex), invariant under $\Pi(t)$ and weakly compact in B_1. By the Tikhonov–Markov theorem the operators $\Pi(t)$ have a fixed point, as we required.

This theorem implies that for each h in a set $\Lambda \subset \mathscr{H}$ of full measure, equation (1_h) has a bounded Besicovitch almost periodic solution. Indeed, let ϕ be an invariant section. We choose a continuous function $\phi_\varepsilon: \mathscr{H} \to B$ for which $\|\phi - \phi_\varepsilon\|_{B_1} \leq \varepsilon$. Then by the ergodic theorem we have

$$\lim_{T \to \infty} \frac{1}{2T} \int_{-T}^{T} \|\phi(h^t) - \phi_\varepsilon(h^t)\|^2 \, dt = \|\phi - \phi_\varepsilon\|_{B_1}{}^2 \leq \varepsilon^2$$

for almost all $h \in H$. The function $\phi_\varepsilon(h^t)$ is almost periodic, and so $\phi(h^t)$ is Besicovitch almost periodic.

The situation regarding the existence of a non-trivial section for homogeneous equations is slightly more complicated.

Definition. We say that $\lambda \in J$ *belongs to the point spectrum of the homogeneous problems* if there is a weakly compact solution of a homogeneous equation of the form (1_h) such that the expression

$$\frac{1}{T} \int_0^T \exp(-i\lambda t) u(t) \, dt \quad (T \to +\infty)$$

has a non-zero weak limiting point.

Let $m_\lambda(u)$ denote an arbitrary limiting point of this expression:

Theorem 8. *If λ is a point in the point spectrum, then the equation $\hat{L}u + i\lambda u = 0$ has a non-trivial section.*

Proof. Without loss of generality we may assume that $\lambda = 0$, that is, that some equation $\hat{L}u = 0$ has a weakly compact solution $u(t)$ with a non-zero mean $m_0(u) = a$. Let $\alpha \in B^*$ with $\operatorname{Re} \alpha(a) \neq 0$. We denote by K the closed convex hull of $u(t)$. For each $h \in \mathcal{H}$ we consider the homogeneous equation (1_h); let K_h be the collection of values $\{u(0)\}$ of those solutions of this equation that lie entirely in K. We put

$$g^*(h) = \max_{p \in K_h} \operatorname{Re} \alpha(p), \quad g_*(h) = \min_{p \in K_h} \operatorname{Re} \alpha(p).$$

The functions $g^*(h)$ and $g_*(h)$ are semicontinuous on $\mathcal{H}$, and at least one of them is non-trivial. Indeed, if $g^*(h) = g_*(h)$ almost everywhere on $\mathcal{H}$, then both these functions are continuous almost everywhere, and so they are Riemann integrable. Then, by the Kronecker–Weyl theorem[9] the means $m_0(g^*(h^t))$ and $m_0(g_*(h^t))$ exist for any $h \in \mathcal{H}$ and are equal to zero, but this contradicts the condition $m_0(\operatorname{Re} \alpha(u)) \neq 0$. To be specific, assume that $g^*(h)$ is non-trivial. Let Z stand for the set of measurable mappings $\phi : \mathcal{H} \to B$ for which $\phi(h) \in K_h$ and $\operatorname{Re} \alpha(\phi(h)) = g^*(h)$; it is easy to see that Z is a non-empty, closed, convex set in B_1 that is separated from zero. For the remainder we must proceed as in Theorem 7.

Note that in the periodic case (here $\mathcal{H}$ is a circle) an invariant section is always continuous (more precisely, equivalent to a continuous one); in the almost periodic case this is not so, as is clear from the examples given below.

2. We give a construction of a scalar equation $u' + a(t)u = f(t)$ (a and f are almost periodic functions with an integer two-term basis), which has bounded solutions but not almost periodic ones. For sets Ω and Λ (Ω is introduced in § 3.4) we have $\Omega \neq \mathcal{H}$ and $\Lambda = \mathcal{H}$. Other properties will be mentioned during the course of the construction.

For convenience in reading we give separately a slightly modified version of the classical example of Bohr.

Bohr's example. Let $\mathcal{H}$ be the two-dimensional torus, realised as a square of side 2π, and let λ_1, λ_2 be linearly independent numbers.

[9] See Levitan [73], p. 109.

We put

$$h=\{x_1, x_2\}, \quad h^t=\{x_1+t\lambda_1, x_2+t\lambda_2\}, \quad h_0=\{0, 0\}.$$

We choose numbers $\omega_m = l_m^1\lambda_1 + l_m^2\lambda_2$ (l_m^1 and l_m^2 are integers) so that $m^{2/3} \leqslant 2\omega_m \leqslant 2m^{2/3}$. We set

$$a(h)=\sum \omega_m^{\ 2} \sin(l_m^1 x_1 + l_m^2 x_2),$$
$$a(t)=a(h_0^{\ t})=\sum \omega_m^{\ 2} \sin \omega_m t.$$

Obviously, $a(t)$ is almost periodic. Let $\eta(t)=\int_0^t a(s)\,\mathrm{d}s$. Then

$$\eta(t)=\sum \omega_m(1-\cos \omega_m t)$$
$$=\sum \omega_m \sin^2(\omega_m t/2).$$

By using the inequality $|\sin x| \geqslant |x|/2$ for $|x| \leqslant 1$ we have

$$\eta(t) \geqslant \sum \omega_m \sin^2(\omega_m t/2)$$
$$\geqslant (t^2/4) \sum_{m \geqslant |t|^{3/2}} (1/m^2)$$
$$\geqslant (t^2/4) \int_{|t|^{3/2}-1}^{\infty} \frac{\mathrm{d}s}{s^2}$$
$$=\frac{t^2}{4(|t|^{3/2}-1)},$$

that is, $\lim_{t\to\infty} \eta(t)=+\infty$. The numbers l_m^1 can be taken to be odd; then $a(h)$ is antiperiodic (in the first argument) with period π, that is, $a(x_1+\pi, x_2)=-a(x_1, x_2)$.

Now we come to our construction. We use the notation of Bohr's example. In addition, for $\phi(h)$ $(h \in \mathcal{H})$ we set by definition

$$\frac{\partial}{\partial t}\phi(h)=\lambda_1\frac{\partial \phi}{\partial x_1}+\lambda_2\frac{\partial \phi}{\partial x_2}.$$

Let $a(t)$ be the function in Bohr's example, and $z(t)=\exp(-\int_0^t a(s)\,\mathrm{d}s)$; $z(t)$ satisfies the equation $z'+a(t)z=0$ and decreases with its derivative faster than any power. We take an interval $\Delta \subset J$ of length 2ε small enough that the corresponding portion $d=\{t\lambda_1, t\lambda_2\}=\{h_0^{\ t}\}$ $(t \in \Delta)$ lies strictly inside the square $\{0 \leqslant x_1 \leqslant 2\pi, 0 \leqslant x_2 \leqslant 2\pi\}$. By using a partition of unity for J we can write $z(t)$ in the form

$$z(t)=\sum_{-\infty}^{\infty} \frac{1}{n^2} u_n(t+\varepsilon n),$$

where the $u_n(t)$ are concentrated on Δ and are uniformly bounded together with their derivatives $u_n'(t)$. We put $z_m=\sum_{n=-m}^{m}(1/n^2)u_n(t+\varepsilon n)$. It is clear that $z_m(t)=z(t)$ on some interval $[-T_m, T_m]$ and that $T_m \to +\infty$.

Next, it is not difficult to realise the construction of periodic functions $\psi_n(h)$ with the following properties:

(1) $\psi_n(h)$ and $\partial\psi_n(h)/\partial t$ are continuous everywhere apart from d where they have a discontinuity of the first kind with jumps $u_n(t)$, $u_n'(t)$, respectively;

(2) $\psi_n(h)$ and $\partial\psi_n(h)/\partial t$ are uniformly bounded.

Then we set

$$\phi_m(h) = \sum_{-m}^{m} \frac{1}{n^2} \psi(h^{\varepsilon n}),$$

$$\phi(h) = \sum_{-\infty}^{\infty} \frac{1}{n^2} \psi(h^{\varepsilon n}).$$

It is easy to see that

$$\frac{\partial}{\partial t}\phi = \sum_{-\infty}^{\infty} \frac{1}{n^2} \frac{\partial}{\partial t}\psi(h^{\varepsilon n}) \quad \text{for } h \notin \{h_0{}^t\}. \tag{15}$$

We define $f_m(h)$ and $f(h)$ by

$$f_m(h) = \frac{\partial}{\partial t}\phi_m(h) + a(h)\phi_m(h),$$

$$f(h) = \frac{\partial}{\partial t}\phi(h) + a(h)\phi(h).$$

By construction, $f_m(h)$ is continuous at each point $h \notin \{h_0{}^t\}$. On the compact part of $\{h_0{}^t\}$, where $|t| \leq T_m$, ϕ_m and $\partial\phi_m/\partial t$ have jumps equal to z_m and z_m', respectively. Since $z'(t) + a(t)z(t) = z_m'(t) + a(h_0{}^t)z_m(t) = 0$ for $|t| \leq T_m$, the function $f_m(h)$ does not have discontinuities on this interval. Since the convergence $f_m \to f$ is uniform on $\mathcal{H}$, $f(h)$ is continuous on $\mathcal{H}$.

Thus, for the equation

$$u' + a(h^t)u = f(h^t) \tag{16}$$

the function $\phi(h)$ is an invariant section with discontinuities (which are not removable) only on $\{h_0{}^t\}$. This means that the solution $\phi(h^t)$, where $h \notin \{h_0{}^t\}$, is N-almost periodic but not almost periodic.

We show that (16) does not have almost periodic solutions. Suppose this were not so, then for $h_1 = \{\pi, 0\}$ the equation $u' + a(h_1{}^t)u = f(h_1{}^t)$ would have an almost periodic solution $u(t)$. The difference $\phi(h_1{}^t) - u(t)$ is a non-trivial bounded solution of the homogeneous equation. But since $a(h)$ is antiperiodic with period π, we have $a(h_1{}^t) = -a(t)$, and so the homogeneous equation does not have a non-trivial bounded solution.

Now we prove the stronger assertion: $\phi(h)$ is the unique (essentially) bounded invariant section. By assuming non-uniqueness we have a non-trivial bounded section for the homogeneous equations; we denote it by $\alpha(h)$.

Returning to Bohr's example we set $g(h) = -\sum \omega_m \cos(l_m^1 x_1 + l_m^2 x_2)$. The series $\sum \omega_m^2$ converges and so $g(h)$ is square integrable over $\mathcal{H}$. The relation $g(h^t) - g(h) = \int_0^t a(h^s)\,ds$ (for almost all $h \in \mathcal{H}$) is readily verified by considering partial sums for $a(h)$ and $g(h)$.

The functions $\alpha(h)$ and $\exp(-g(h))$ are non-trivial sections for one and the same scalar equation $u' + a(h^t)u = 0$; consequently, their ratio $\alpha(h)\exp(g(h))$ is invariant under translation on $\mathcal{H}$. Since translations are strictly ergodic, this ratio is constant almost everywhere on $\mathcal{H}$. We find that $\exp(-g(h))$ is essentially bounded.

Let $g = g^+ - g^-$ be the decomposition into positive and negative parts. From $\exp(-g) \geq 1 - g$ it follows that g^- is bounded, while the antiperiodicity of g implies that g^+ is bounded.

Thus, we have found that $\exp(-g)$ and $\exp(g)$ are bounded. Then there is a point $h_2 \in \mathcal{H}$ for which $u(t) = \exp(-g(h_2{}^t))$ and $1/u(t)$ are bounded. This means that the equation $u' + a(h_2{}^t)u = 0$ has a bounded solution that is separated from zero. But then $u_t' + a(h_0{}^t)u = 0$ also must have such a solution, which contradicts the original properties of Bohr's example. This completes the construction of the required example. We can also give for the homogeneous equation an example of an invariant section that is not equivalent to a continuous one. For this we must consider the equation $u_t' + ia(h^t)u = 0$, where $a(h)$ is the function in Bohr's example. The required section is defined by $\phi(h) = \exp(-ig(h))$.

Comments and references to the literature

§ **2.** The minimax method was given by Favard [106] in the finite-dimensional case and in the case of a uniformly convex space by Amerio (see Amerio & Prouse [2]). Both Favard and Amerio assumed the condition of (two-sided) separation and not of semiseparation, and they studied almost periodic solutions and not N-almost periodic ones. The general formulation of the 'minimax' Theorem 3 and also the example of a non-compact solution is due to the authors. For finite-dimensional spaces the condition of semiseparation implies that of separation (see Chapter 7, § 4). This is not so in the infinite-dimensional case. As an example we can take the semigroup of right translations in $\mathcal{L}^2(0, \infty)$.

§§ **3 and 4.** Lemma 2 (in a more general form) is in Gel'fand's article [36]. The other results are taken from the works of Zhikov. Note that the problem of completely non-distal extensions and related questions of the existence of N-almost periodic solutions (see § 3, Corollary 1) are open.

§ **5.** Questions of compactness occupy a central place in the investigations of Amerio, who started from the well-known works of Sobolev [102] on the homogeneous wave equation. He succeeded in studying the non-homogeneous wave equation, and in proving the theorem on the integral (see Chapter 6, § 1) and certain abstract results about compactness within the framework of the minimax method (all these results are contained, for example, in Theorem 5). Our approach to questions of compactness differs from Amerio's. The important argument about a point of strong continuity is given in an article by Kadets [63], where there is a proof of Corollary 2. (This argument was used independently by Zhikov [50], p. 184, in connection with the criterion for a point spectrum.) For a unitary group this criterion can be proved on the basis of a spectral resolution (see Lax & Phillips [69], p. 139).

§ **6.** Example 3 is in the fundamental article of Bochner & von Neumann [29], who discuss the case of compact solutions. The Bochner–von Neumann method is based on the rather complicated techniques of generalised harmonic analysis. The original formulation of Theorem 6 (with a derivation of the results of Bochner & von Neumann) is given in an article of Zhikov [47]. K. V. Valikov (private communication) and Perov and Ta Kuang Khai [97] have made some simplifications to the original proof.

§ **7.** The problem of constructing a linear equation solvable in the class of bounded functions but not in the class of almost periodic functions was discussed in the original article of Favard [106]. But Favard only succeeded in showing (on the basis of Bohr's example) that the separation property does not carry over automatically to the limiting equations.

9 The method of monotonic operators

1 General properties of monotonic operators

1. Let V be a separable reflexive Banach space, V^* be its dual, and (y, x) denote the value of $y \in V^*$at $x \in V$.

Definition. An operator $A: V \to V^*$ is called *monotonic* if

$$\operatorname{Re}(Ax_1 - Ax_2, x_1 - x_2) \geqslant 0 \quad (x_1, x_2 \in V).$$

In what follows, to simplify the notation we assume that V is real, so that the symbol Re in the definition can be dropped. An operator A is called *bounded* if it carries bounded sets in V into bounded sets in V^*.

Definition. An operator $A: V \to V^*$ is called *semicontinuous* if the scalar function $t \to (A(u+tv), w)$ is continuous for any $u, v, w \in V$.

Lemma 1. *Let $A: V \to V^*$ be a monotonic, semicontinuous operator. An element $u \in V$ satisfies the equation $Au = f$ if and only if*

$$(Av - f, v - u) \geqslant 0 \tag{1}$$

for any $v \in V$.

Proof. Let $Au = f$. Then we have

$$(Av - f, v - u) = (Av - Au, v - u) \geqslant 0.$$

Conversely, if in (1) we take $v = u + tw$ $(t > 0, w \in V)$, then we obtain

$$(A(u + tw) - f, tw) \geqslant 0.$$

After cancelling by t and letting t tend to 0 we obtain $(Au - f, w) \geqslant 0$ for any $w \in V$. Hence it is clear that $Au = f$.

***Remark* 1.** We have assumed that the operator A is defined on the whole of V, but it is clear that Lemma 1 is still valid when the domain $D(A)$ of A is a linear manifold dense in V; it is merely necessary to assume, in addition, that u belongs to $D(A)$.

The next proposition is a consequence of Lemma 1.

Proposition 1. *For a monotonic semicontinuous operator* $A: V \to V^*$ *the set of solutions of the equation* $Au = f$ *is closed and convex.*

For a proof it is enough to note that if two elements u_1, u_2 satisfy (1), then $(u_1+u_2)/2$ also satisfies (1).

Proposition 2. *A monotonic semicontinuous bounded operator* $A: V \to V^*$ *gives a continuous mapping if* V *is endowed with the strong topology and* V^* *with the weak.*[1]

Proof. Let $u_n \to u$ and $f_n = A(u_n)$. The sequence $\{f_n\}$ is bounded, and consequently it has a weak limit point, which we denote by y. We can take the limit in the inequality $(Av - f_n, v - u_n) \geqslant 0$, and as a result obtain $Au = y$, as we required.

Theorem 1. *Suppose that a monotonic semicontinuous bounded operator* $A: V \to V^*$ *is such that*

$$(Au, u) \geqslant 0 \quad (\|u\| = r_0). \tag{2}$$

Then there is a $u \in V$ *with* $\|u\| \leqslant r_0$, *such that* $Au = 0$.

Proof. Let V_n be an expanding sequence of finite-dimensional subspaces of V whose union is dense in V, and let δ_n denote the restriction operator of $v \in V^*$ to V_n. We consider the continuous finite-dimensional operators

$$A_n = \delta_n A : V_n \to V_n{}^*.$$

Since $(A_n u, u) \geqslant 0$ $(\|u\| = r_0, u \in V_n)$, we obtain from the finite-dimensional result (see Proposition 3 below) $u_n \in V_n$ with $\|u_n\| \leqslant r_0$ such that $A_n u_n = 0$. We have the inequality

$$(A_n v, v - u_n) = (Av, v - u_n) \geqslant 0, \tag{3}$$

where $v \in V_m$, $m \leqslant n$. The bounded sequence $\{u_n\}$ contains a weak limit point which we denote by u. We take the limit as $n \to \infty$ in (3) for a fixed $v \in V_m$; as a result we obtain $(Av, v - u) \geqslant 0$ $(v \in V_m)$. Then

[1] This property is called *demicontinuity.*

by continuity (using Proposition 2) this inequality turns out to hold for any $v \in V$. It remains to apply Lemma 1.

Theorem 2. *Suppose that a monotonic semicontinuous bounded operator $A: V \to V^*$ has the coercive property*

$$\lim_{\|u\| \to \infty} \frac{(Au, u)}{\|u\|} = +\infty.$$

Then for any $f \in V^*$ the equation $Au = f$ has at least one solution.

For a proof it is sufficient to recognise that $Au - f$ satisfies (2) for a suitable choice of r_0.

Proposition 3. *Let $P: R^m \to R^m$ be a continuous mapping of the Euclidean space R^m into itself such that $(P(\xi), \xi) \geqslant 0$ for any ξ in the sphere $\|\xi\| = r_0$. Then there is a ξ, $\|\xi\| \leqslant r_0$, such that $P(\xi) = 0$.*

Proof. If $P(\xi) \neq 0$ on the ball $K = \{\xi : \|\xi\| \leqslant r_0\}$, then we can consider the operator

$$\xi \to -r_0 \frac{P(\xi)}{\|P(\xi)\|} : K \to K,$$

which in this case is continuous. Then from the classical fixed point theorem of Bohl–Brouwer it follows that that there is a ξ such that

$$\xi = \frac{-r_0 P(\xi)}{\|P(\xi)\|}. \tag{4}$$

But then $|\xi| = r_0$, and by multiplying both sides of (4) scalarly by $P(\xi)$ we obtain

$$(P(\xi), \xi) = -r_0 \|P(\xi)\| < 0,$$

which is impossible.[2]

2. So far we have assumed that the monotonic operator is defined on the whole of the space V. For an application to evolution problems we require to study the problem of the solvability of an equation of the form

$$Lu \equiv \Lambda u + Au = f, \tag{5}$$

where A is, as before, a monotonic semicontinuous bounded operator $V \to V^*$, and Λ is an unbounded linear operator $D(\Lambda) \to V^*$ (here

[2] To use Proposition 3 in the proof of Theorem 1 we need to equip the space V_n with a scalar product. This enables us to identify V_n^* with V_n.

$D(\Lambda) \subset V$ and $D(\Lambda)$ is dense in V) subject to the following conditions:

(1) Λ is closed and monotonic, that is, $(\Lambda u, u) \geq 0$ for all $u \in D(\Lambda)$;

(2) there is a sequence of expanding subspaces $V_n \subset D(\Lambda)$ whose union is dense in V and which are such that $\Lambda: V_n \to V^*$ is bounded for each n;

(3) if δ_n is the restriction operator of the functional $v \in V^*$ to V_n, then there is a constant l such that

$$\|\Lambda u\| \leq l \|\delta_n \Lambda u\| \quad (u \in V_n).$$

Theorem 3. *Suppose that, in addition to conditions* (1)–(3), *the operator A is coercive. Then* (5) *has at least one solution for any* $f \in V^*$.

Proof. The operator $Q: D(\Lambda) \to V^*$, where $Qu \equiv \Lambda u + Au - f$, is monotonic. There is a number r_0 (depending on $\|f\|$) such that $(Qu, u) \geq 0$ $(u \in D(\Lambda), \|u\| = r_0)$. The operator $Q_n = \delta_n q: V_n \to V_n{}^*$ satisfies the conditions of Theorem 1. Let $Q_n u_n = 0$, $\|u_n\| \leq r_0$. Since the operators $\delta_n A$ are uniformly bounded, the sequence $\delta_n \Lambda u_n$, and so the sequence Λu_n, is bounded. Hence, since Λ is a closed operator, there is at least one $u \in d(\Lambda)$ for which $u_{n_k} \rightharpoonup u$ and $\Lambda u_{n_k} \rightharpoonup \Lambda u$. Then it is necessary to proceed as in Theorem 1 (using Remark 1). This proves Theorem 3.

3. Now we consider the question of the fixed points of non-expansive operators.

Let H be a Hilbert space. An operator $T: H \to H$ is called *non-expansive* if $\|Tx_1 - Tx_2\| \leq \|x_1 - x_2\|$. Here there is an important connection with monotonicity since the operator $A = I - T$ is monotonic. In fact

$$(Ax_1 - Ax_2, x_1 - x_2) = (x_1 - x_2, x_1 - x_2) - (Tx_1 - Tx_2, x_1 - x_2)$$
$$\geq \|x_1 - x_2\|^2 - \|Tx_1 - Tx_2\| \|x_1 - x_2\| \geq 0.$$

Theorem 4. *A semigroup* $U(t)$ $(t \geq 0)$ *of non-linear non-expansive operators has a common fixed point if and only if it is bounded, that is,*

$$\sup_{t \geq 0} \|U(t)x\| < \infty \quad (x \in H).$$

Similarly, a non-expansive operator T has a fixed point if and only if the semigroup of powers T^m $(m = 1, 2, \ldots)$ *is bounded.*

Proof. The necessity is obvious; we prove the sufficiency. First we study the single operator case. We fix an $x_0 \in H$ and set

$$K_0 = \{T^n x_0, n \geq 0\}, \quad K_m = \{T^n x_0, n \geq m\}.$$

Then we choose a positive number $d > \operatorname{diam} K_0$ and denote by $S^d(x)$ the ball in H of radius d and with centre at x. We define

$$\Omega_m = \bigcap_{x \in K_m} S^d(x), \quad \Omega = \bigcup_{m \geq 0} \Omega_m.$$

It is obvious that Ω_m is convex, $\Omega_m \subset \Omega_{m+1}$, and the set Ω is bounded. Since T is a non-expansive operator we obtain

$$T\Omega_m \subset \Omega_{m+1}, \quad T\Omega \subset \Omega. \tag{6}$$

We introduce a family of operators T_ε defined by

$$T_\varepsilon x = \varepsilon(Tx - Tx_0) + Tx_0 \quad (0 < \varepsilon < 1).$$

Since $x_0 \in \Omega_0$ and the sets Ω_m are convex, taking (6) into account we obtain $T_\varepsilon \Omega \subset \Omega$. For $\varepsilon < 1$ the operator T_ε satisfies the contraction mapping principle; we denote its fixed point by x_ε. The boundedness of the set $\{x_\varepsilon\}$ follows from (6) and the boundedness of Ω; without loss of generality we may assume that $x_\varepsilon \rightharpoonup x$ (weakly) as $\varepsilon \to 1$. Setting $A_\varepsilon = I - T_\varepsilon$ we have $A_\varepsilon x_\varepsilon = 0$. Taking the limit in $(A_\varepsilon y, y - x_\varepsilon) \geq 0$ $(y \in H)$ we conclude that $Ax = 0$ and $Tx = x$. Thus, the operator T has a closed convex set of fixed points.

Now we turn to the semigroup $U(t)$. The operator $T = U(t_0)$ satisfies the boundedness condition; let N_{t_0} be the closed convex set of its fixed points. From the commutativity of the operators $U(t)$ $(t \geq 0)$ it follows that N_{t_0} is an invariant set, that is, $U(t)N_{t_0} \subset N_{t_0}$ $(t \geq 0)$. Let N be a minimal (relative to inclusion) closed convex invariant set. It can be proved by the preceding argument that the operator $T = U(t_0)$ has at least one fixed point in N, and so it follows that N consists of the common fixed points of the operators $U(t_0)$.

2 Solvability of the Cauchy problem for an evolution equation

Let E be a separable reflexive embedded space; this means that there is a Hilbert space H such that the embedding $E \subset H \subset E^*$ is dense and continuous, and that the bilinear form (y, x) $(y \in E^*, x \in E)$ coincides with the scalar product on H whenever $x, y \in H$. We shall use the following notation for norms: $\|x\|$ is an E-norm, $\|y\|_*$ is an E^*-norm, and $|x|$ is an H-norm.

Our aim is a study of the evolution operator $\mathrm{d}/\mathrm{d}t + A$, but first we shall be concerned with constructing the differentiation operator $\mathrm{d}/\mathrm{d}t$.

1. Let $[0, T]$ be a finite interval, S be the set of all trigonometric polynomials with coefficients in E, $p > 1$, and $p^{-1} + q^{-1} = 1$. We define D to be the closure of the set S with respect to the norm

$$\|u\|_D = \left(\int_0^T \|u\|^p \,\mathrm{d}t\right)^{1/p} + \left(\int_0^T \|u'\|_*^q \,\mathrm{d}t\right)^{1/q}$$
$$= \|u\|_{\mathscr{L}^p(0,T;E)} + \|u'\|_{\mathscr{L}^q(0,T;E^*)}.$$

The following important proposition holds.

Proposition 4. *Each element in D is an H-continuous function on* $[0, T]$, *so that it is meaningful to talk about the initial condition* $u|_{t=0} = u(0) \in H$. *Moreover, for each* $u_0 \in H$ *there is at least one* $u(t) \in D$ *such that* $u(0) = u_0$.

Proof. Suppose that $u \in S$. Since $|u| \leqslant \|u\|$, by the theorem about the mean there is a point $t_0 = t_0(u)$ for which

$$|u(t_0)| = T^{-1}\left(\int_0^T |u(t)|^p \,\mathrm{d}t\right)^{1/p} \leqslant T^{-1}\|u\|_D.$$

Therefore, from the formula $|u(t)|^2 = |u(t_0)|^2 + 2\int_{t_0}^T (u', u)\,\mathrm{d}\eta$, and by using the obvious inequality

$$\int_0^T |(u', u)|\,\mathrm{d}t \leqslant \left(\int \|u'\|_*{}^q\right)^{1/q}\left(\int_0^T \|u\|^p \,\mathrm{d}t\right\}^{1/p}$$
$$\leqslant \|u\|_D{}^2$$

we obtain

$$\sup_{0\leqslant t\leqslant T} |u(t)|^2 \leqslant (T^{-1} + 2)\|u\|_D{}^2.$$

It follows from here that the elements in D are H-continuous functions. Then, by an uncomplicated construction, for any $u_0 \in H$ we can find an extension $u(t) \in D$, and Proposition 4 is proved.

We put

$$V = \mathscr{L}^p(0, T; E), \quad V^* = \mathscr{L}^q(0, T; E^*).$$

We define the duality between V and V^* by the usual formula: the value $\langle v, u\rangle$ of $v \in V^*$ at $u \in V$ is given by $\langle v, u\rangle = \int_0^T (v, u)\,\mathrm{d}t$.

We consider the operator $\Lambda = \mathrm{d}/\mathrm{d}t$ with the domain $\{u \in D: u(0) = 0\}$, and shall prove that it satisfies conditions (1)–(3) in § 1.

The operator Λ is monotonic since

$$\begin{aligned}2\langle Lu, u\rangle &= 2\int_0^T (u', u)\,\mathrm{d}t\\ &= |u(T)|^2 - |u(0)|^2\\ &= |u(T)|^2 \geq 0.\end{aligned}$$

Because it is obvious that Λ is a closed operator, we only need to give the construction of the subspaces V_n. For V_n we take the collection of all trigonometric polynomials

$$u(t) = \sum_{k=1}^{n} b_k \sin(k\pi t/T) \quad (b_k \in E).$$

It is well known that every element $z \in V$ is the sum of its Fourier sine series, and if z_n denotes the nth partial sum of this series, then $\|z_n\|_V \leq l\|z\|_V$. For $u \in V_n$ we have

$$\langle \Lambda u, z\rangle = \int_0^T (u', z)\,\mathrm{d}t = \int_0^T (u', z_n)\,\mathrm{d}t.$$

Therefore,

$$\begin{aligned}\|\Lambda u\| &= \sup_{\|z\|_V=1} \langle \Lambda u, z\rangle \leq \sup_{\|z_n\|\leq l} \langle \Lambda u, z_n\rangle\\ &= l\|\delta_n \Lambda u\|,\end{aligned}$$

that is, condition (3) is fulfilled.

2. We impose the following conditions on an operator A:

(1) $A: E \to E^*$ is monotonic and semicontinuous;

(2) there are constants $c_i > 0$ $(i = 1, 2, 3, 4)$ and a $p > 1$ such that

$$(Ax, x) \geq c_1\|x\|^p - c_2, \tag{7}$$

$$\|Ax\|_* \leq c_3\|x\|^{p-1} + c_4. \tag{8}$$

The first of these estimates means that A is coercive, and the second that it is bounded.

Now we must interpret A as an operator from V into V^*. With this aim we note that $A: E \to E^*$ sends strongly convergent sequences into weakly convergent ones. Then, for a function with values in a separable space (for instance, E and E^*), strong and weak measurability coincide. Hence it follows that the function $Au(t)$ is E^*-measurable provided $u(t)$ is E-measurable. It is clear from (8) that

A acts from V into V^* since

$$\int_0^T \|Au\|_*^{\,q}\,dt \leqslant c_3 \int_0^T \|u\|^{(p-1)q}\,dt + c_4 T$$
$$= c_3 \int_0^T \|u\|^p\,dt + c_4 T.$$

It is not difficult also to verify that $A: V \to V^*$ is monotonic and semicontinuous.

In what follows we shall be interested not only in the operator A but also an operator $B: V \to V^*$ of the form $Bu = A(\xi + u)$, where ξ is a fixed element in V. We prove the coercive estimate

$$\langle Bu, u\rangle = \int_0^T (A(\xi+u), u)\,dt$$
$$\geqslant k_1 \|u\|_V^{\,p} - k_2$$

with positive constants k_1 and k_2 depending on $\|\xi\|_V$ and T. Indeed, we have

$$\int_0^T (A(\xi+u), u)\,dt = \int_0^T (A(\xi+u), \xi+u)\,dt$$
$$- \int_0^T (A(\xi+u), \xi)\,dt.$$

We estimate the first term on the right from below using (7) and the second from above with the help of (8). As a result we obtain:

$$\langle Bu, u\rangle \geqslant c_1 \int_0^T \|\xi+u\|^p\,dt - c_2 T - c_3$$
$$\times \int_0^T \|\xi+u\|^{p-1}\|\xi\|dt - c_4 \int_0^T \|\xi\|\,dt,$$

$$\langle Bu, u\rangle \geqslant c_1 \int_0^T \|\xi+u\|^p dt - c_3$$
$$\times \left(\int_0^T \|\xi+u\|^p\,dt\right)^{1/q} \left(\int_0^T \|\xi\|^p\,dt\right)^{1/p} - c_5,$$

$$\langle Bu, u\rangle \geqslant c_1 \|\xi+u\|_V^{\,p} - c_3 \|\xi+u\|_V^{\,p/q} \|\xi\|_V - c_5. \tag{9}$$

Since $p/q < p$, the required estimate follows from the last inequality.

3. We consider the following initial-value problem:

$$Lu \equiv u_t' + Au = f(t) \quad (0 \leqslant t \leqslant T),$$
$$u(0) = u_0 \in H,\ f(t) \in \mathcal{L}^q(0, T; E^*).$$

Theorem 5. *For any $u_0 \in H$ and $f \in \mathscr{L}^q(0, T; E^*)$ there is a unique H-continuous function $u(t)$ such that $u(0) = u_0$, $u \in \mathscr{L}^p(0, T; E)$, $u' \in \mathscr{L}^q(0, T; E^*)$, and the equation $u' + Au = f$ is satisfied almost everywhere on $[0, T]$.*

Proof. For an initial value $u_0 \in H$ we take an extension $\xi(t) \in D$ and put $u(t) = \xi(t) + z(t)$. Then $z(t)$ must satisfy the equation

$$z' + A(\xi(t) + z) = f(t) - \xi'(t) = g(t),$$
$$z(0) = 0.$$

We write this equation as $\Lambda z + Bz = g$. We verified earlier that the operators Λ and B satisfy the corresponding conditions in Theorem 3. Only the uniqueness remains. If $u_1(t)$ and $u_2(t)$ are two solutions of (9), then $(u_1 - u_2)' + Au_1 - Au_2 = 0$. By scalar multiplying this relation by $u_1 - u_2$ and integrating we obtain

$$|u_1(t) - u_2(t)|^2 - |u_1(0) - u_2(0)|^2 + 2 \int_0^t (Au_1 - Au_2, u_1 - u_2)\, d\tau = 0.$$

Since $|u_1(0) - u_2(0)|^2 = 0$ and the integral term is non–negative we have $u_1(t) \equiv u_2(t)$, as we required.

3 The evolution equation on the entire line: questions of the boundedness and the compactness of solutions

We denote by X the space of measurable functions $f: J \to E^*$ with the norm

$$\|f\|_X = \sup_{t \in J} \left(\int_0^1 \|f(t+s)\|_*^q \, ds \right)^{1/q}.$$

Let X_{comp} denote the subspace of X consisting of those f for which the family of translates $\{f(t+\tau)\}$ $(\tau \in J)$ is compact in $\mathscr{L}^q(-T, T; E^*)$, and $\mathring{X}$ denote the subspace of those $f \in X$ for which the family of translates $\{f(t+\tau)\}$ $(\tau \in J)$ is compact in X. Obviously, $\mathring{X} \subset X_{\text{comp}} \subset X$.

1. The next lemma establishes the boundedness (in the H-norm) of solutions on the half-line $[t_0, \infty)$.

Lemma 2. *Let $f \in X$. Then any solution $u(t)$ $(t \geqslant t_0)$ of the equation $Lu = f$ has the estimate*

$$\sup_{t \geqslant t_0} |u(t)| \leqslant \max\{|u(t_0)|, l\}, \tag{10}$$

where the constant l does not depend on the solution.

Proof. We scalar multiply the equality $u'+Au=f$ by $u(t)$ and integrate from s to t $(s\leqslant t\leqslant t_0)$:

$$\tfrac{1}{2}|u(t)|^2-\tfrac{1}{2}|u(s)|^2+\int_s^t (Au, u)\,\mathrm{d}\eta=\int_s^t (f, u)\,\mathrm{d}\eta.$$

By applying the coercive estimate (7) we obtain

$$\begin{aligned}\tfrac{1}{2}|u(t)|^2-\tfrac{1}{2}|u(s)|^2+c_1\int_s^t \|u\|^p\,\mathrm{d}\eta\\ &\leqslant c_2(t-s)+\int_s^t \|f\|_*\|u\|\,\mathrm{d}\eta\\ &\leqslant c_2(t-s)+\left(\int_s^t \|f\|_*{}^q\,\mathrm{d}\eta\right)^{1/q}\left(\int_s^t \|u\|^p\,\mathrm{d}\eta\right)^{1/p}.\end{aligned}$$

We estimate the last term on the right by using the inequality $|ab|\leqslant p^{-1}\varepsilon^p a^p+q^{-1}\varepsilon^{-q}b^q$; as a result we obtain

$$|u(t)|^2+c_1\int_s^t \|u\|^p\,\mathrm{d}\eta\leqslant|u(s)|^2+c_0\left\{(t-s)+\left(\int_s^t \|f\|_*{}^q\,\mathrm{d}\eta\right)^{1/q}\right\}. \quad (11)$$

12226

Below we shall refer to the boundedness of a number of quantities; we shall have in mind boundedness by constants not depending on solutions. We assume that the half-line $[t_0, \infty)$ contains an interval $\Delta=[t_1-1, t_1]$ such that

$$\Gamma=|u(t_1)|=\max_{t\in\Delta}|u(t)|.$$

Then from (11) follows the boundedness of $\int_\Delta \|u\|^p\,\mathrm{d}\eta$, and consequently, of $\int_\Delta |u|^p\,\mathrm{d}\eta$. Then we can find a t_2 in Δ such that the norm $|u(t_2)|$ is bounded. Now if in (11) we take $s=t_2$ and $t=t_1$, then we conclude that Γ is bounded by some constant l_1. Hence, as is easily seen, we obtain the inequality

$$\sup_{t\geqslant t_0}|u(t)|\leqslant\max\{\max_{t_0\leqslant t\leqslant t_0+1}|u(t)|, l_1\}.$$

By estimating $\max_{t_0\leqslant t\leqslant t_0+1}|u(t)|$ in terms of $|u(t_0)|$ with the help of (11) we obtain the required inequality. This proves the lemma.

From (11) we obtain

$$\sup_{t\geqslant t_0+T}\int_0^T \|u(t+s)\|^p\,\mathrm{d}s,\ \sup_{t\geqslant t_0+T}\int_0^T \|u'(t+s)\|_*{}^q\,\mathrm{d}s\leqslant C(|u(t_0)|, T). \quad (12)$$

We obtain some other inequalities. Let $f_1, f_2\in X$, and $u_1(t)$, $u_2(t)$ $(t\geqslant t_0)$ be solutions of the equations $Lu=f_1$, $Lu=f_2$, respectively.

We have

$$\tfrac{1}{2}|u_1(t)-u_2(t)|^2-\tfrac{1}{2}|u_1(t_0)-u_2(t_0)|^2 + \int_{t_0}^{t}(Au_1-Au_2, u_1-u_2)\,dt = \int_{t_0}^{t}(f_1-f_2, u_1-u_2)\,dt.$$

By using monotonicity and then (12) we find that

$$|u_1(t)-u_2(t)|^2 \leq |u_1(t_0)-u_2(t_0)|^2 + 2\left(\int_{t_0}^{t}\|f_1-f_2\|_*{}^q\,dt\right)^{1/q}\left(\int_{t_0}^{t}\|u_1-u_2\|^p\,dt\right)^{1/p},$$

$$|u_1(t)-u_2(t)|^2 \leq |u_1(t_0)-u_2(t_0)|^2 + \left(\int_{t_0}^{t}\|f_1-f_2\|_*{}^q\,dt\right)^{1/q} C(|u_1(0)|, |u_2(0)|, \|f_1\|_X, \|f_2\|_X). \tag{13}$$

2. In this subsection we digress from parabolic evolution equations and prove certain abstract propositions about compactness. Suppose that we have a triple of continuously embedded separable Banach spaces $B_1 \subset B \subset B_2$, where the embedding $B_1 \subset B$ is compact and the space $B_2{}^*$ is dense in B^*.

For certain $p \geq 1$ and $p_1 > 1$ we set

$$\Pi = \{u \in \mathscr{L}^p(0, T; B_1): u \in C(0, T; B), u' \in \mathscr{L}^{p_1}(0, T; B_2)\},$$
$$|u|_\Pi = \|u\|_{\mathscr{L}^p(0,T;B_1)} + \|u\|_{C(0,T;B)} + \|u'\|_{\mathscr{L}^{p_1}(0,T;B_2)}.$$

Lemma 3. *The embedding $\Pi \to \mathscr{L}^p(0, T; B)$ is compact.*

As a preliminary we prove the following assertion.

Proposition 5. *Suppose that the embedding $B_1 \subset B$ is compact, and that there is a family $\{u_\alpha(t)\}$ $(0 \leq t \leq T)$ for which we postulate*

(a) boundedness in $\mathscr{L}^p(0, T; B_1)$ and $C(0, T; B)$;

(b) weak equicontinuity in B.

Then this family is compact in $\mathscr{L}^p(0, T; B)$.

Proof. Without loss of generality we may assume that there is a sequence of finite-dimensional operators $I_m: B \to B$ such that $I_m x \to Ix = x$ $(x \in B)$. Indeed, we can replace the space B by the universal space $C(0, 1)$ while preserving the conditions of Proposition 5, and such a sequence exists in $C(0, 1)$. Since the strong convergence of linear operators is always uniform on compact sets,

$$\|I_m x - x\| \leq \varepsilon \|x\|_{B_1} \quad (x \in B_1, m \geq m(\varepsilon)).$$

By putting $u_\alpha^m = I_m u_\alpha$ we obtain

$$\left(\int_0^T \|u_\alpha - u_\alpha^m\|^p \,dt\right)^{1/p} \leqslant \varepsilon\left(\int_0^T \|u_\alpha\|_{B_1}{}^p \,dt\right)^{1/p} \leqslant \varepsilon l,$$

that is, the set $\{I_m u_\alpha(t)\}$ is an εl-net for $\{u_\alpha\}$. But the compactness of the set $\{I_m u_\alpha\}$ follows easily from conditions (a) and (b), and Proposition 5 is proved.

To prove Lemma 3 we verify that the unit ball in Π satisfies condition (b). But for $g \in B_2^*$ we have

$$\begin{aligned}|(u(t), g)|_{t_0}^{t_0+\Delta} &\leqslant \int_{t_0}^{t_0+\Delta} |(u', g)| \,dt \\ &\leqslant \int_{t_0}^{t_0+\Delta} \|u'\|_{B_1}\|g\|_{B_2^*} \,dt \\ &\leqslant \|g\|_{B_2^*}\left(\int_{t_0}^{t_0+\Delta} \|u'\|^p \,dt\right)^{1/p} \Delta^{1/q} \\ &\leqslant \|g\|_{B_2^*}\Delta^{1/q}.\end{aligned}$$

From this (b) follows if we take into account the fact that the ball that is being considered is bounded in $C(0, T; B)$, and the assumption that the space B_2^* is dense in B^*.

3. Now we explain when the equation $Lu = f$ has a solution bounded on the whole line, and when this solution is compact. With this aim we assume, in addition, that the embedding $E \subset H$ is compact (the 'compactness condition').

Lemma 4. *For any $f \in X$ the equation $Lu = f$ has at least one solution that is H-bounded on the whole line. If, additionally, $f \in X_{\text{comp}}$, then a bounded solution is compact.*

Proof. Let $u_n(t)$ $(t \geqslant -n)$ be a solution with the initial condition $u|_{t=-n} = 0$. In view of the estimates (10) and (12) and Lemma 3, the sequence $\{u_n(t)\}$ is compact in $\mathscr{L}^p(-T, T; H\}$. Without loss of generality we may assume that $|u_m(t) - u_n(t)| \to 0$ almost everywhere on J. But since $|u_m(t) - u_n(t)| \leqslant |u_m(t_0) - u_n(t_0)|$ $(t \geqslant t_0)$, the sequence $\{u_n(t)\}$ is in fact fundamental in the sense of local convergence of J. If $u(t)$ $(t \in J)$ is a limit function, then it must be a solution of our equation. For let $z(t)$ $(t \geqslant t_0)$ be a solution with the initial condition $z(t_0) = u(t_0)$. Since $|u_n(t) - z(t)| \leqslant |u_n(t_0) - z(t_0)| \to 0$, we have $u(t) = z(t)$ $(t \geqslant t_0)$.

We turn to the question of compactness. To be specific we consider a solution $u(t)$ defined on the half-line $t \geq t_0$. For an arbitrary sequence of numbers $t_m \to \infty$ the sequence $\{u(t+t_m)\}$ is fundamental in $\mathcal{L}^p(-T, T; H)$. Without loss of generality we may assume the convergence $u(t+t_m) \to \hat{u}(t)$ almost everywhere on J, and that $f(t+t_m) \to \hat{f}(t)$ in the sense of $\mathcal{L}^q(-T, T; E^*)$. But then from (13) we obtain (by analogy with the preceding) that $\hat{u}(t)$ must be a solution of the limiting equation $Lu = \hat{f}$ and that the convergence $u(t+t_m) \overset{H}{\to} \hat{u}(t)$ is uniform on intervals. This proves the lemma.

4 Almost periodic solutions of the evolution equation

1. Suppose that the free term $f(t)$ is in $\mathring{X}$, that is, $f(t)$ is an almost periodic (in the sense of Stepanov) function $J \to E^*$. We are interested in solutions that are Bohr almost periodic functions $J \to H$, and denote the class of such solutions by $\mathring{C}(H)$.

Let $\mathfrak{M}$ be the module of Fourier exponents of $f(t)$.

Theorem 6. *If the compactness condition holds, then the equation $Lu = f$ has at least one solution $u \in \mathring{C}(H)$ with a module of Fourier exponents belonging to $\mathfrak{M}$.*

Note that when $f(t) \equiv \bar{f}$, the almost periodic solution in Theorem 6 also does not depend on t, we obtain the solvability of the stationary equation $Au = \bar{f}$. We only need to emphasise that Theorem 6 is based essentially on the compactness condition in contrast to a stationary problem and evolution problems on a finite interval. At the end of this section we discuss the situation when the compactness condition is not assumed.

Proof of Theorem **6.** Let $\mathcal{H}$ denote the closure of the family $\{f(s+t)\}$ $(t \in J)$ in X, and for $\hat{f}(s) \in \mathcal{H}$ we set $\hat{f}^t = \hat{f}(s+t)$.

It follows from the monotonicity of the operator A that if $u_1(t)$ and $u_2(t)$ $(t \geq t_0)$ are solutions of the equation $Lu = \hat{f}$, then the norm $|u_1(t) - u_2(t)|$ is non-increasing on $[t_0, \infty)$. But then, as was shown in Chapter 7, § 4, any two solutions compact on the entire line must satisfy the invariance identity

$$|u_1(t) - u_2(t)| \equiv \text{const.} \quad (t \in J). \tag{14}$$

Now we give another proof of this identity that does not depend on the special results in Chapter 7.

Let $u_1(t)$ and $u_2(t)$ $(t \in J)$ be fixed compact solutions of an equation of the form $Lu = \hat{f}$. We consider in the space $H \times \mathcal{H}$ the natural semigroup connected with the equation $Lu=\hat{f}$.[3] We select a compact invariant subset $Z\subset H\times\mathscr{H}$ containing the fixed trajectories $\{u_1(t),\hat{f}^t\}$ and $\{u_2(t),\hat{f}^t\}$ (Z can be the union of the closures of these trajectories). Let P denote the fibre over $\hat{f}\in\mathscr{H}$; thus, $P\subset H$ is a collection of initial values $\{u(0)\}$ of those solutions of $Lu=\hat{f}(\hat{f}\in\mathscr{H}$ is fixed) such that the corresponding trajectory $\{u(t),\hat{f}^t\}$ belongs to Z. For an element $\hat{f}$ we fix a 'returning' sequence $t_m\to-\infty$: $f(t+t_m)\overset{X}{\to}f(t)$.

For each $u(0)\in P$ we fix (in view of possible non-uniqueness) a compact solution $u(t)$ such that $\{u(t),\hat{f}^t\}\subset Z$. We introduce operators $T_m:P\to H$, where by definition $T_m u(0)=u(t_m)$. It is obvious that the sequence of operators T_m is compact in the topology of pointwise convergence; let T be some limit point. By the definition of pointwise convergence, for any $v_1(0)$, $v_2(0)\in P$ we have

$$p_1=Tv_1(0)=\lim_{m\to\infty} v_1(t'_m),\quad p_2=Tv_2(0)=\lim_{m\to\infty} v_2(t'_m),$$

where $\{t'_m\}$ is some subsequence of $\{t_m\}$. Since the solutions $v_1(t)$ and $v_2(t)$ are compact and uniformly continuous on the entire line, we can assume the local convergence

$$v_1(t+t'_m)\xrightarrow{\text{loc}} p_1(t),\quad v_2(t+t'_m)\xrightarrow{\text{loc}} p_2(t).$$

From the invariance of the set Z it follows at once that $p_1(0), p_2(0)\in P$. Since the norm $|v_1(t)-v_2(t)|$ is a non-increasing function of $t\in J$,

$$|p_1(t)-p_2(t)|=\lim_{m\to\infty}|v_1(t+t'_m)-v_2(t+t'_m)|\equiv\text{const}.$$

In other words, we have proved (14) for any $p_1(0)$, $p_2(0)\in TP$. Consequently, it remains to prove that $TP=P$. Since T is the limit of non-contraction operators T_m, it is itself non-contracting, that is, $|Tp_1-Tp_2|\geqslant|p_1-p_2|$. Hence it is enough to refer to Proposition 1 in Chapter 1.

Now we prove that the half-sum of compact solutions is also a solution. We put $v(t_0)=(u_1(t_0)+u_2(t_0))/2$ and $\alpha=|u_1(0)-u_2(0)|$. Suppose that $v(t)\,(t\geqslant t_0)$ is a solution with the initial condition $v|_{t=t_0}=v(t_0)$.

From the inequalities

$$|u_i(t)-v(t)|\leqslant|u_i(t_0)-v(t_0)|=\alpha/2\quad(i=1,2;\ t\geqslant t_0),$$

[3] The important 'continuity condition' (see Chapter 7, § 1) follows easily from (13).

the identity $|u_1(t)-u_2(t)|\equiv\alpha$, and the strict convexity of the norm in a Hilbert space, we immediately obtain

$$v(t)=(u_1(t)+u_2(t))/2 \quad (t\geqslant t_0).$$

Then by letting $t_0\to-\infty$ we see that the half-sum $(u_1(t)+u_2(t))/2$ is a solution.

Thus, the set of compact solutions of an equation of the form $Lu=\hat{f}$ is convex. This makes the situation close enough to the linear case (see Chapter 8, § 2). An almost periodic solution $\mathring{u}(t)$ can be selected by the minimax method:

$$\sup_{t\in J}|\mathring{u}(t)|=\inf\sup_{t\in J}|u(t)|,$$

where inf is taken over all the compact solutions of the equation $Lu=f$. This proves Theorem 6.

2. Now we turn to a discussion of this theorem.

The proof of Theorem 6 (after Lemma 4) essentially is not based on special properties of a parabolic evolution equation. Therefore it is not difficult to give Theorem 6 a completely general form.

We consider in a Banach space B a general non-linear equation

$$u'=F(u,t). \tag{15}$$

We assume that the right-hand side is almost periodic, and the 'continuity condition' (for the details see Chapter 7). Let $\rho=\rho(p,q)$ be a function that is symmetric and continuous on $B\times B$, and vanishes only on the diagonal. Suppose that ρ satisfies the strict triangle inequality, that is, $\rho(p_1,q)+\rho(q,p_2)=\rho(p_1,p_2)$ implies that q belongs to the interval $[p_1,p_2]$.

Theorem 7. *Suppose that the solving operators of the equations $u'=\hat{F}(u,t)$ are ρ-non-expansive. Then if equation* (15) *has a compact solution for $t\geqslant 0$, then it has at least one almost periodic solution.*

Now we consider the parabolic evolution equation without the condition that the embedding $E\subset H$ is compact. Lemma 4 is a key result; without the compactness condition one cannot succeed in proving the existence of a solution bounded on the whole line, not referring to compactness.

Let μ be a normalised invariant measure on $\mathcal{H}$. It turns out that for almost all $h=\hat{f}\in\mathcal{H}$ the corresponding equation $Lu=\hat{f}$ has a solution bounded on the entire line; moreover, these solutions are bounded by a common constant and are Besicovitch almost periodic.

To prove this let H_1 be the space of μ-measurable mappings $\phi : \mathscr{H} \to H$ with the norm $|\phi|_1^2 = \int_{\mathscr{H}} |\phi(h)|^2 \, d\mu$. Let $S_h(t)$ ($t \geq 0$, $h \in \mathscr{H}$) be the solving operators for the equation $Lu = \hat{f}$. We introduce a family of operators $\Pi(t): H_1 \to H_1$ ($t \geq 0$) by

$$\Pi(t)\phi(h) = S_{h^{-t}}\phi(h^{-t}).$$

It is easy to see that these operators commute, that is, they form a semigroup. Next, they are H_1-non-expansive. For since the operators $S_h(t): H \to H$ are non-expansive we have

$$|\Pi(t)\phi_1 - \Pi(t)\phi_2|_1^2 \leq \int_{\mathscr{H}} |\phi_1(h^{-t}) - \phi_2(h^{-t})|^2 \, d\mu$$

$$= \int_{\mathscr{H}} |\phi_1(h) - \phi_2(h)|^2 \, d\mu = |\phi_1 - \phi_2|_1^2.$$

Consider the estimate (10); from it we obtain immediately the inequality

$$|\Pi(t)\phi(h)|^2 \leq |\phi(h^{-t})|^2 + l^2.$$

Therefore, $|\Pi(t)\phi|_1^2 \leq |\phi|_1^2 + l^2$, that is, the semigroup $\Pi(t)$ is bounded. If $\psi(h)$ is a common fixed point, then as in the linear case (see Chapter 8, § 7), for almost all $h \in \mathscr{H}$ the solutions $u(t) = \psi(h^t)$ are Besicovitch almost periodic, and the following limit exists:

$$\lim_{T\to\infty} \frac{1}{2T} \int_{-T}^{T} |\psi(h^t)|^2 \, dt = |\psi|_1^2. \tag{16}$$

The solutions $u(t) = \psi(h^t)$ are bounded by a common constant; indeed, (16) and the inequality $|u(t)| \geq |u(t_0)| - l$ ($t \leq t_0$) implies that $\sup_{t\in J} |u(t)|^2 \leq l^2 + |\psi|_1^2$.

Finally, we mention a simple result that does not require the compactness condition. We retain the estimates (7) and (8) for the operator A but replace the monotonicity condition by the stronger condition:

$$(Ax - Ay, x - y) \geq \gamma |x - y|^2 \quad (x, y \in E)$$

with the constant $\gamma > 0$. We prove that the equation $Lu = f$ has a unique almost periodic solution. For two solutions $u_1(t)$ and $u_2(t)$ ($t \geq t_0$) we have

$$\frac{1}{2}\frac{d}{dt}|u_1 - u_2|^2 + |Au_1 - Au_2| = 0,$$

$$\frac{d}{dt}|u_1 - u_2|^2 + \gamma |u_1 - u_2|^2 \leq 0,$$

$$|u_1(t) - u_2(t)| \leq \exp(-\gamma(t - t_0))|u_1(t_0) - u_2(t_0)|.$$

Therefore, if $U(t, t_0)$ $(t \geq t_0)$ is a solving operator $u(t_0) \to u(t)$, then $|U(t, t_0)p - U(t, t_0)q| \leq \exp(-\gamma(t-t_0))|p-q|$. We consider a family of operators $\Pi(t): \mathring{C}(H) \to \mathring{C}(H)$, where

$$\Pi(t)\phi(s) = U(t+s, s)\phi(s-t).$$

The operators $\Pi(t)$ commute, and for $t > 0$ they are contractions; consequently, they have a common fixed point, which is obviously an almost periodic solution.

Comments and references to the literature

§§ 1 and 2. The material in these sections is mainly classical for that subject area of non-linear analysis conventionally called 'the method of monotonic operators'. A valuable feature of this method is its application to non-linear elliptic and parabolic equations. Although these applications are very important in their own right and served as the source of the method of monotonic operators, we have not touched on them (as regards this question, see the extensive literature, for example, the book by Lions [79]). The fixed point theorem (proved by Zhikov) is a generalisation of the standard theorem (usually attributed to Browder): a non-expansive operator $T: H \to H$ has a fixed point if it leaves a bounded convex set invariant.

§ 3. Lemma 3 and Proposition 5 are versions of fairly well known 'compactness lemmas' (Lions, [79], Russian p. 70).

§ 4. The results here are extracted from Zhikov's article [52]. It should be noted that we have restricted ourselves only to the results about almost periodic functions that are proved most easily. A more complete theory embracing, in particular, non-linear equations of hyperbolic type, is presented in the article of Zhikov & Levitan [57].

10 Linear equations in a Banach space (questions of admissibility and dichotomy)

Notation

We use the notation: $C = C(B)$ is the space of continuous functions $J \to B$ with the sup norm; $C^+ = C^+(B)$ and $C^- = C^-(B)$ are analogous spaces for the half-lines $J^+ = [0, \infty)$ and $J^- = (-\infty, 0]$, respectively; $M^p = M^p(B)$ and $\mathscr{L}^p = \mathscr{L}^p(-\infty, \infty; B)$ $(p \geq 1)$ are the spaces of measurable functions $u : J \to B$ with the norms

$$\sup_{t \in J} \left(\int_0^1 \|u(t+s)\|^p \, \mathrm{d}s \right)^{1/p} \quad \text{and} \quad \left(\int_{-\infty}^{\infty} \|u\|^p \, \mathrm{d}t \right)^{1/p},$$

respectively; the constants $c_0, c_1, \ldots, l_0, l_1, \ldots$ are assumed positive.

1 Preliminary results

1. Our basic object is a linear operator $L \equiv \mathrm{d}/\mathrm{d}t + A(t)$, where $A(t)$ $(t \in J)$ is generally an unbounded operator in a Banach space B. We assume the following *condition of right-solvability*: for any initial value $u(t_0) \in B$, the equation $Lu = 0$ has a unique strongly continuous solution $u(t)$ $(t \geq t_0)$. We assume also that the solving operators $U(t, t_0)$ $(t \geq t_0)$ are strongly continuous with respect to $t \geq t_0$ and satisfy the estimate

$$\|U(t, t_0)\| \leq l_0 \exp(c_0(t - t_0)) \quad (t \geq t_0). \tag{1}$$

The solution $u(t)$ $(t \geq t_0)$ of the non-homogeneous equation $Lu = f$ $(f \in M^1(B))$ is determined from the usual formula

$$u(t) = U(t, t_0)u(t_0) + \int_{t_0}^{t} U(t, s)f(s) \, \mathrm{d}s.$$

If $u(t)$ $(t \in J)$ satisfies this equality for any $t \geq t_0 \in J$, we speak of the solution on the entire line, and write $Lu = f$ $(t \in J)$. We obtain the following estimate immediately from (1):

$$\|u(t)\| \leqslant c_0^* \left\{ \|u(t_0)\| + \int_{t_0}^{t_0+1} \|f(s)\| \, ds \right\} \quad (t_0 \leqslant t \leqslant t_0+1). \tag{2}$$

We mention some properties of the solutions. Let $Lu = f$ $(t > t_0)$ and let $\phi(t)$ $(t \in J)$ be a smooth scalar function with support on (t_0, ∞). We set $v = \phi u$ $(t \in J)$. Then $Lv = \phi f + \phi' u$ $(t \in J)$. In what follows we always take ϕ to be a smooth mapping $J \to [0, 1]$; when ϕ needs to be prescribed more exactly, we give its support (supp ϕ) and the set on which $\phi(t) = 1$.

Now we define L as an unbounded operator in the space $C = C(B)$. For this it is enough to specify its domain $D = \{u \in C : Lu \in C\}$. Similarly for the spaces $M^p(B)$ and $\mathscr{L}^p(-\infty, \infty; B)$. From (2) it follows that in all cases we have the embedding $D \subset C$. It is perfectly clear that these operators are closed.

An operator L is called *regular* if the equation $Lu = f$ is uniquely solvable in C for any $f \in C$. From Banach's theorem we obtain the existence of a bounded inverse $L^{-1} : C \to C$ and the estimate

$$\|u\|_C \leqslant k \|Lu\|_C. \tag{3}$$

L is called *correct* if (3) holds for any u with $Lu \in C$, and *strongly correct* if $\|u\|_C \leqslant k \|Lu\|_{M^2}$ for any $u \in C$ with $Lu \in M^2$. Since $C \subset M^2$, strong correctness implies correctness.

A pair (W, F) of Banach function spaces is called *admissible* for an operator L if $Lu = f$ has at least one solution $u \in W$ for any $f \in F$. From Banach's open mapping theorem we obtain that for any $f \in F$ there is a solution $u \in W$ such that

$$\|u\|_W \leqslant k \|f\|_F \tag{4}$$

with a constant k not depending on $f \in F$. If (C, C) is an admissible pair, then L is called *weakly regular*.

It turns out that the concept of regularity is closely linked with an important concept in stability theory: the exponential splitting or the exponential dichotomy. Suppose that we have a variable decomposition $B = N_1(t_0) + N_2(t_0)$ into a direct sum of subspaces, and let $P_1(t_0)$ and $P_2(t_0)$ be the corresponding projection operators. Also let $u(t)$ $(t \geqslant t_0)$ be a solution of $Lu = 0$. The operator L has the *exponential dichotomy* on the whole line J if

(d_1) $U(t, t_0) N_i(t_0) \subset N_i(t) \quad (t \geqslant t_0, i = 1, 2)$,

(d_2) $\sup_{t_0 \in J} \|P_i(t_0)\| < \infty$,

(d$_3$) $\|u(t)\| \leqslant l_1\|u(t_0)\| \exp(-c_1(t-t_0)) \quad (t \geqslant t_0,\, u(t_0) \in N_1(t_0))$,

(d$_4$) for every initial value $u(t_0) \in N_2(t_0)$ there is a unique extension to the whole line such that

$$\|u(t)\| \leqslant l_1\|u(t_0)\| \exp(c_1(t-t_0)) \quad (t \leqslant t_0).$$

Among other things, when the estimate (1) holds condition d$_2$ is fulfilled automatically. To prove this we fix a sufficiently large $\Delta > 0$ that

$$l_1^{-1} \exp(c_1\Delta) - l_1 \exp(-c_1\Delta) = \alpha > 0.$$

Let $u_i(t_0) \in N_i(t_0)$ with $\|u_i(t_0)\| = 1$. For $t = t_0 + \Delta$ we have

$$\|u_1(t)\| \leqslant l_1 \exp(-c\Delta), \quad \|u_2(t)\| \geqslant l_1^{-1} \exp(c_1\Delta),$$

which together with (1) gives

$$\begin{aligned}\|u(t_0)\| &= \|u_1(t_0) + u_2(t_0)\| \\ &\geqslant l_0^{-1} \exp(-c_0\Delta)\|u_1(t) + u_2(t)\| \\ &\geqslant l_0^{-1} \exp(-c_0\Delta)\{\|u_2(t)\| - \|u_1(t)\|\} \\ &\geqslant \alpha_0 l_0^{-1} \exp(-c_0\Delta) \\ &= \gamma > 0,\end{aligned}$$

that is, $\|P_i(t_0)\| \leqslant \gamma^{-1}$.

Now note that from the exponential dichotomy on J it follows that L is regular; the proof consists of a straightforward construction of the Green's function. For this, by using condition (d$_4$), we introduce the bounded operators $\Omega(t, t_0) = U(t, t_0)P_2(t_0)$ for any $t, t_0 \in J$. Further, we set

$$G(t, s) = \begin{cases} U(t, s)P_1(s) & \text{for } t \geqslant s, \\ -\Omega(t, s) & \text{for } t \leqslant s. \end{cases} \tag{5}$$

Conditions (d$_2$)–(d$_4$) imply the estimate

$$\|G(t, s)\| \leqslant l \exp(-c_1|t-s|). \tag{6}$$

Therefore, the function $u(t) = \int_J G(t, s)f(s)\, ds$ is bounded. Then it is easy to verify the identity $U(t, t_0)G(t_0, s) = G(t, s) - \chi_{t_0}^t(s)U(t, s)$, where $\chi_{t_0}^t(s)$ is the characteristic function of the interval (t_0, t). By applying this identity to $f(s)$ and integrating, we see that $u(t)$ is a solution of the non-homogeneous equation. The uniqueness of the solution in C follows from conditions (d$_2$)–(d$_4$).

Next we formulate the concept of exponential dichotomy on the half-line J^-. Let N_2 be a complemented subspace in B, N_1 be complementary to N_2, and $u(t)$ $(t \geqslant t_0)$ be a solution of the homogeneous equation. We say that the exponential dichotomy holds on J^- if

(d_1^-) for any $u(0)\in N_2$ there exists a unique extension to J^- such that

$$\|u(t)\|\leq l_1\|u(t_0)\| \exp(c_1(t-t_0)) \quad (t\leq t_0\leq 0);$$

(d_2^-) a solution $u(t)$ $(t_0\leq t\leq 0)$ satisfies the estimate

$$\|u(t)\|\leq l_1\|u(t_0)\| \exp(-c_1(t-t_0)) \quad (t_0\leq t\leq 0)$$

provided that $u(0)\in N_1$.

We construct a Green's function. For $t_0\leq 0$ we set by definition

$$N_2(t_0)=\{u(t_0)\in B : u(0)\in N_2\},$$
$$N_1(t_0)=\{u(t_0)\in B : u(0)\in N_1\},$$

where in the first we take $u(t)$ to be the solution in condition (d_1^-), and in the second $u(t)$ is arbitrary. It follows from condition (d_1^-) that $N_1(t_0)\cap N_2(t_0)=\{0\}$. We show that these subspaces are mutually complementary. Suppose that $x(t_0)\in B$ and $x(t)$ $(t\geq t_0)$ is the corresponding solution of the homogeneous equation. We set $y(0)=P_1(0)x(0)$ and $z(0)=P_2(0)x(0)$. By assumption, for $z(0)$ there is a unique extension $z(t)$ to J with the estimate in (d_1^-); therefore $z(t_0)\in N_2(t_0)$. Since $y(t_0)=x(t_0)-z(t_0)\in N_1(t_0)$, we have proved that the subspaces are mutually complementary.

The subsequent arguments are similar to the case of the whole line. We introduce the operator-function $\Omega(t,s)=U(t,s)P_2(t,s)$ $(t,s\in J^-)$. The Green's function $G(t,s)$ $(t,s\in J^-)$ is given by (5), and satisfies the exponential estimate. Therefore, the function $u(t)=\int_{-\infty}^{0} G(t,s)f(s)\,ds$ is a solution in C^-, that is, the pair (C^-, C^-) is admissible.

2. Now we study the concept of correctness in more detail. Let L be a correct operator and $u(t)$ $(a\leq t\leq b)$ a solution of $Lu=f$. Suppose that $T>0$ is such that the interval $[t_0-2T, t_0+2T]$ lies in (a, b).

We choose a smooth function $\phi : J\to[0,1]$ for which

$$\phi(t)=1 \text{ for } t\in[t_0-T, t_0+T],$$
$$\operatorname{supp}\phi\subset[t_0-2T, t_0+2T], \quad |\phi'|\leq 2/T.$$

For $v=\phi u$ we have $Lv=f\phi+\phi' u$ $(t\in J)$. Therefore, from (2) we obtain

$$\sup_{|t-t_0|\leq T}\|u(t)\|\leq k\sup_{|t-t_0|\leq 2T}\|f(t)\|+\frac{2k}{T}\sup_{[a,b]}\|u(t)\|. \tag{7}$$

Before using this estimate let us note that the notions of local convergence, local closure, etc. can be naturally introduced in the spaces C, M^p, and others. For instance, we say that $f_m \to_{\text{loc}} f$ in M^p if the sequence $\{f_m\}$ is bounded in M^p, and $\|\phi(f_m - f)\|_{M^p} \to 0$ for any smooth ϕ with compact support.

It turns out that the range of a correct operator is locally closed in C. For suppose that $Lu_m = f_m$, where $u_m, f_m \in C$, and that $f_m \to_{\text{loc}} f$. If we apply (7) to the difference $u_m - u_n$ we find that the sequence $\{u_m\}$ is locally fundamental in C. The limit function is a solution of the equation $Lu = f$.

Similarly it can be proved that the range of a strongly correct operator is locally closed in $M^2(B)$.

2 The connection between regularity and the exponential dichotomy on the whole line

In this section our aim is to prove the following assertion.

Theorem 1. *Regularity is equivalent to the exponential dichotomy on J.*

In one direction this assertion has already been proved in § 1; it remains to prove that regularity implies exponential dichotomy on J. Our proof is constructed so that regularity is really only used in the last stage, until then our arguments are based only on correctness.

Suppose that $u(t)$ $(t \geq t_0)$ is a solution of $Lu = 0$. We call it *uniformly stable to the right* if

$$\|u(t)\| \leq l\|u(s)\| \quad (t \geq s \geq t_0) \tag{8}$$

and *uniformly exponentially stable to the right* if

$$\|u(t)\| \leq l_1 \exp(-c_1(t-s)) \quad (t \geq s \geq t_0). \tag{9}$$

In a similar way we speak of uniform stability and exponential uniform stability to the left if a solution $u(t)$ is defined for $t \leq t_0$.

Lemma 1. *If L is correct, then uniform stability (to the right or left) implies uniform exponential stability (to the right or left), and the constants l_1 and c_1 depend only on l in* (8) *and k in* (3).

Proof. To be definite we consider stability to the right. We show that we can find a number $T = T(l, k)$ such that

$$\|u(s+t)\| \leq \|u(s)\|/2 \quad (t \geq T, s \geq t_0). \tag{10}$$

For suppose there is a solution $z(t)$ with the estimate (8) and an interval $\Delta = [a, b] \subset [t_0, \infty)$ such that $\|z(a)\| = 1$ and $\|z(b)\| \geq \frac{1}{2}$. From

(8) we obtain immediately

$$\|z(t)\| \leqslant l, \quad \|z(t)\| \geqslant 1/2l \quad (t \in \Delta). \tag{11}$$

Let ϕ be such that supp $\phi \subset \Delta$ and $\phi(t) = 1$ for $t \in [a+\varepsilon, b-\varepsilon]$, and let

$$f = \phi(t) z(t) \|z(t)\|^{-1},$$
$$u(t) = z(t) \int_{-\infty}^{t} \phi(s) \|z(s)\|^{-1} \, ds.$$

By using (11) it is easy to calculate that $\|u\|_C \geqslant 2^{-1} l^{-2} (\Delta - 2\varepsilon)$. Since $\|f\|_C = 1$ and $Lu = f$, from (3) we obtain that $\Delta \leqslant 2kl^2$, which proves (10). Then this estimate and (8) imply (11) if we take $l_1 = 2l$ and $c_1 = T^{-1} \ln 2$. Thus, Lemma 1 is proved.

We introduce the following manifolds:

$$N_1(t_0) = \{u(t_0) \in B : Lu = 0, \sup_{t \geqslant t_0} \|u(t)\| < \infty\}, \tag{12}$$

$$N_2(t_0) = \{u(t_0) \in B : Lu = 0, \sup_{t \leqslant t_0} \|u(t)\| < \infty\}. \tag{13}$$

First of all we note that under our conditions the solution of the homogeneous equation can have an infinite number of extensions to the whole line. In (13) it is required that such an extension is bounded on $(-\infty, t_0)$; correctness implies the uniqueness of this extension. We prove the inequalities

$$\|u(t)\| \leqslant l \|u(t_0)\| \quad (t \geqslant t_0, u(t_0) \in N_1(t_0)), \tag{14}$$

$$\|u(t)\| \leqslant l \|u(t_0)\| \quad (t \leqslant t_0, u(t_0) \in N_2(t_0)). \tag{15}$$

To this end let ϕ be such that:

$$\text{supp } \phi \subset [t_0, \infty), \phi(t) = 1$$
$$(t \geqslant t_0 + 1), |\phi'(t)| \leqslant 2. \tag{16}$$

If $u(t_0) \in N_1(t_0)$, then for $v = \phi u$ we have $Lv = \phi' u = f$ $(t \in J)$. Therefore, from (3) and (1) we obtain

$$\sup_{t \geqslant t_0 + 1} \|u(t)\| \leqslant \|v\|_C \leqslant k \|f\|_C \leqslant 2k \exp c_0 \|u(t_0)\|,$$

which together with (1) gives (14). Next, if $u(t_0) \in N_2(t_0)$, then by extending the solution to the whole line and setting $v = (1 - \phi) u$, we obtain (15) by a similar method.

The estimates (14) and (15) indicate that the manifolds $N_1(t_0)$ and $N_2(t_0)$ are closed; by Lemma 1, these estimates imply the corresponding exponential estimates. Consequently, it remains to prove that the subspaces $N_1(t_0)$ and $N_2(t_0)$ are mutually complementary.

Let $x(t_0)$ be arbitrary and $x(t)$ $(t \geqslant t_0)$ be a solution of the homogeneous equation. We put $f = \phi' x(t)$ $(t \in J)$, where ϕ satisfies the conditions (16). By the regularity property, the equation $Lu = f$ has a unique solution $u \in C$; the function $v(t) = \phi(t)x(t)$ $(t \in J)$ also satisfies this equation. We set $z(t) = v(t) - u(t)$ and $y(t) = x(t)z(t)$. Then obviously $Lz = 0$ $(t \in J)$ and $Ly = 0$ $(t \geqslant t_0)$. It is clear from the definition of ϕ that

$$\sup_{t \leqslant t_0} \|z(t)\| < \infty, \qquad \sup_{t \geqslant t_0 + 1} \|y(t)\| < \infty,$$

that is, $z(t_0) \in N_2(t_0)$ and $y(t_0) \in N_1(t_0)$. Thus we have obtained the decomposition $x(t_0) = y(t_0) + z(t_0)$, which proves that $N_1(t_0)$ and $N_2(t_0)$ are mutually complementary. Theorem 1 is proved.

We consider a family of operators L_α satisfying the inequality (1) uniformly with respect to α. We say that the L_α are *uniformly regular* if the inverse operators $L_\alpha^{-1} : C \to C$ are uniformly bounded. Furthermore, we say that the L_α have the uniform exponential dichotomy on J if the constants in conditions (d_2)–(d_4) are bounded above and separated from zero.

The preceding arguments prove that uniform regularity is equivalent to the uniform exponential dichotomy on J.

Finally, we remark that a regular operator is automatically strongly correct. In fact, we can prove more than this, namely, the existence of a bounded inverse $L^{-1} : M^1(B) \to C(B)$. If $f \in M^1(B)$, then we set $u(t) = \int_{-\infty}^{\infty} G(t, s) f(s)\, ds$. In view of (6) we have

$$\|u(t)\| \leqslant l \int_{-\infty}^{\infty} \exp(-c_1|s|) \|f(s+t)\|\, ds.$$

By dividing the axis of integration into unit intervals we find that

$$\begin{aligned} \|u(t)\| &\leqslant 2l \sup_{t \in J} \left(\int_0^1 \|f(s+t)\|\, ds \right) \sum_{m=0}^{\infty} \exp(-cm) \\ &= k_1 \|f\|_{M^1}. \end{aligned} \tag{17}$$

3 Theorems on regularity

In many cases it is desirable to deduce regularity from simpler properties, such as correctness or weak regularity. We prefer to start from correctness since firstly it arises in a number of problems on stability, and secondly, weak regularity can be reduced to correctness by duality arguments.

Below we distinguish two basic situations in which correctness implies regularity.

1. We say that the *compactness condition* holds if the operators $U(t, t_0)$ $(t > t_0)$ are compact. The simplest examples show that the compactness condition is not enough, we need certain conditions on the dependence of $A(t)$ on $t \in J$.

A function $\xi : J \to R$ (R is a metric space) is called *Poisson stable* (*has the P-property*) if $\xi(t + t_m) \to \xi(t)$ $(t \in J)$ for some sequence $t_m \to -\infty$.

If the operator $A(t)$ were bounded, then the P-property would mean the Poisson stability of the function $A(t): J \to \text{Hom}\,(B, B)$. In general, the P-property must be formulated in terms of a solving operator, namely, there exists a sequence $t_m \to -\infty$ such that

$$U(t + t_m, s + t_m)x \to U(t, s)x \quad (t \geqslant s, x \in B).$$

Theorem 2. *If the compactness condition holds, then a correct operator having the P-property is regular.*

Proof. We prove that the exponential dichotomy on the half-line J^- holds. To this end we consider the subspace $N_2 = N_2(0)$ defined by (13). The compactness condition implies that this subspace is finite-dimensional; suppose N_1 is a complement of it. Since property (d_1^-) is already ensured by the estimate (15) and Lemma 1, we prove property (d_2^-).

As a preliminary we prove uniform stability to the right, that is, the inequality

$$\|u(t)\| \leqslant l\|u(s)\| \quad (t_0 \leqslant s \leqslant t \leqslant 0, u(0) \in N_1). \tag{18}$$

By assuming otherwise we find a sequence of solutions u_m with $\sup \|u_m\| = 1$ and $\|u_m(a_m)\| \to 0$ $(a_m \leqslant t \leqslant 0)$. It follows from the compactness condition that the set $\{u_m(t)\}$ is compact for any $t \leqslant 0$. Hence, if $z(t)$ is a limit point of the sequence $u_m(t)$, then $z(t) \in C^-$, $Lz = 0$, and $z(0) \in N_1$, that is, $z \equiv 0$ and it follows that $u_m(t) \to_{\text{loc}} 0$. But then the norm $\|u_m(t)\|$ reaches its maximum value (unity) at a point s_m such that $s_m \to -\infty$ and $a_m - s_m \to -\infty$. We apply the estimate (7) to $u_m(t)$, setting $(a, b) = (a_m, 0)$, $t_0 = s_m$, and $T = \min\,\{|s_m|, |a_m - s_m|\}$. As a result we obtain a contradiction which proves (18), and together with it the exponential dichotomy on J^-, and consequently, the admissibility of the pair (C^-, C^-). But then, for every $f \in C^-$ the equation $Lu = f$ has a solution with the estimate $\|u\|_{C^-} \leqslant k_1 \|f\|_{C^-}$.

For each $f \in C$ we put $f_m(t) = f(t - t_m)$, and suppose that $v_m(t)$ $(t \in J^-)$ is a solution of $Lv = f_m$ with the above estimate. Then we set $u_m(t) = v_m(t + t_m)$.

Thus, we have

$$Lu_m = f\ (t \leqslant -t_m), \quad \sup_{t \leqslant -t_m} \|u_m(t)\| \leqslant k_1 \|f\|_C.$$

The difference $z_{m,n} = u_m - u_n$ satisfies the homogeneous equation on the half-line $(-\infty, b_{m,n})$, where $b_{m,n} = \min\ (-t_m, -t_n)$. From (7) we easily obtain $\sup_{t \leqslant \frac{1}{2} b_{m,n}} \|z_m(t)\| \to 0$, that is, in any case the sequence $\{u_m\}$ is locally fundamental. By taking the limit we obtain a bounded solution of $Lu = f$. This proves Theorem 2.

2. As well as the operator L we consider the dual operator L^* formally given by $L^* = -\mathrm{d}/\mathrm{d}t + A^*(t)$. Regarding L^* (since it is referred to in general) we assume the condition of left-solvability and an estimate of the form (1). As before, we denote the value of $y \in B^*$ at $x \in B$ by (y, x).

An informal determination that the operators L and L^* are duals is contained in Green's formula. Namely, it is assumed that if $Lu = f\ (t \in J)$ and $L^*v = g\ (t \in J)$ for $f \in M^2(B)$ and $g \in M^2(B^*)$, then

$$(v, u)|_{t_1}^{t_2} = \int_{t_1}^{t_2} [(g, u) - (v, f)]\ \mathrm{d}t.$$

Theorem 3. *If L is strongly correct and L^* is correct, then they are both regular.*

Proof. We define L and L^* as unbounded operators in the spaces $B_1 = \mathscr{L}^2(-\infty, \infty; B)$ and $B_1{}^* = \mathscr{L}^2(-\infty, \infty; B^*)$. It follows from inequalities of the form (2) that $\|u(t)\| \to 0$ and $\|v(t)\| \to \infty$ as $t \to \infty$ for $u \in D(L)$ and $v \in D(L^*)$. Therefore, Green's formula becomes

$$\int_J (v, Lu)\ \mathrm{d}t = \int_J (L^*v, u)\ \mathrm{d}t$$
$$(u \in D(L),\ v \in D(L^*)).$$

Hence it follows that the operator adjoint to L (relative to the natural duality between B_1 and $B_1{}^*$) is some extension of L^*. To prove the coincidence of these operators it is enough to establish that L^* has a non-empty resolvent set. With this in mind we consider the operator $L^* - \lambda I$, where $\lambda > 0$ is sufficiently large. Since the solutions of the homogeneous equation $L^*v - \lambda v = 0$ decrease exponentially as $t \to -\infty$, this operator is regular. Consequently, it is sufficient to prove the invertibility in $B_1{}^*$ of a regular operator. To this end we take a function $f \in B_1{}^*$ with compact support and put $v = \int_J G(t, s) f(s)\ \mathrm{d}s$, where G is the Green's function of a regular operator. The solution v decreases exponentially and satisfies the estimate

$$\|v(t)\| \leq l\left\{\int_{-\infty}^{t} \exp(-c_1(t-s))\|f(s)\|\,\mathrm{d}s + \int_{t}^{\infty} \exp(c_1(t-s))\|f(s)\|\,\mathrm{d}s\right\} = l\{\psi_1(t)+\psi_2(t)\}.$$

Since $\psi_i' + c_1\psi_i = \|f\|$, after multiplying by ψ_i and integrating we have

$$c_1\int_J |\psi_i|^2\,\mathrm{d}t = \int_J \psi_i\|f\|\,\mathrm{d}t \leq \left(\int_J |\psi_i|^2\,\mathrm{d}t\right)^{1/2}\left(\int_J \|f\|^2\,\mathrm{d}t\right)^{1/2},$$

that is, $\|v\|_{B_{1^*}} \leq l/2c_1\|f\|_{B_{1^*}}$, which (in view of closedness) gives invertibility in B_{1^*}.

Now we assume that L is not regular. Then we can find a function $f_0 \in C$ with compact support such that $Lu = f_0$ has no solutions in C. We prove that the range of L (as of an operator in B_1) is not dense in B_1. Assuming otherwise, we find a sequence $u_m \in B_1$ such that $Lu_m = f_m \to f_0$. Since $u_m \in C$ and $\|f_m - f_0\|_{M^2} \leq \|f_m - f_0\|_{B_1}$, from strong correctness we obtain $\|u_m - u_0\|_C \to 0$, that is, the equation $Lu = f_0$ is solvable in C. Hence, the range of L is not dense in B_1.

Since L^* is adjoint to L, the orthogonal complement of the range of L consists of the zeros of L^*. These are the elements of $C(B^*)$ that contradict the correctness of L^*. This proves that L is regular.

From what has been proved above the regular operator L is invertible in B_1. By a theorem of Phillips (Yosida [61], p. 273), a conjugate operator L^* is invertible in B_{1^*}. This, together with correctness, gives regularity, and Theorem 3 is proved.

3. It would be useful to establish the duality of the concepts of correctness and weak regularity. In this direction we have only been able to obtain the following result.

Lemma 2. *If one of L or L* is weakly regular, then the other is correct.*

Proof. We prove that L^* is correct if L is weakly regular. Let $\eta(t)$ be an odd continuous scalar function such that $\eta(t) = 1$ for $t \geq 1$. We fix an $\varepsilon > 0$ sufficiently small that the pair (C, C) remains admissible for the family of operators $\mathrm{d}/\mathrm{d}t + A(t+s) - \varepsilon\eta(t)I$ with a common constant k in an equality of the form (4).[1]

Assuming that L^* is not correct, we can find sequences $\{y_m\}$, $\{g_m\} \subset C(B^*)$ for which $L^* y_m = g_m$, $\|y_m\|_C = 1$ and $\|g_m\|_C \to 0$. We select a $t_m \in J$ such that $\|y_m(t_m)\| \geqslant \frac{1}{2}$ and put $v_m(t) = y_m(t+t_m)$, $\tilde{g}_m(t) = g(t+t_m)$, and $L_m = \mathrm{d}/\mathrm{d}t + A(t+t_m)$.

From $\|v_m(0)\| \geqslant \frac{1}{2}$ and an estimate of the form (2) we obtain immediately

$$\|v_m(t)\| \geqslant \alpha > 0 \quad (0 \leqslant t \leqslant 1,\ m \geqslant m_0). \tag{19}$$

Then putting

$$v_m^{\varepsilon}(t) = v_m(t) \exp\left(-\varepsilon \int_0^t \eta(s)\,\mathrm{d}s\right),$$

$$\tilde{g}_m^{\varepsilon} = \tilde{g}_m(t) \exp\left(-\varepsilon \int_0^t \eta(s)\,\mathrm{d}s\right),$$

and bearing in mind that $v_m^{\varepsilon}(t) \to 0$ as $t \to \infty$ and $L_m{}^* v_m^{\varepsilon} - \varepsilon\eta v_m^{\varepsilon} = \tilde{g}_m^{\varepsilon}$, by Green's formula we have

$$\int_J (f, v_m^{\varepsilon})\,\mathrm{d}t = \int_J (u, \tilde{g}_m^{\varepsilon})\,\mathrm{d}t, \tag{20}$$

where f is an arbitrary element in C, and u is a solution of $L_m u - \varepsilon\eta u = f$ satisfying an estimate of the form (4).

We consider the expression $\lambda_m(f) = \int_J (f, v_m^{\varepsilon})\,\mathrm{d}t$ as a functional over C. Since $\|\lambda_m\| = \int_J \|v_m^{\varepsilon}\|\,\mathrm{d}t$, then as follows from (19), $\|\lambda_m\| \geqslant \beta > 0$. On the other hand, from (20) we have

$$\|\lambda_m\| \leqslant k \int_J \|\tilde{g}_m^{\varepsilon}\|\,\mathrm{d}t \to 0.$$

In a similar way, from the admissibility of (C, C) for L^* we can obtain the correctness of L.

4 Examples

1. Suppose that the space B is finite-dimensional, and $A(t): J \to \mathrm{Hom}\,(B, B)$ is a bounded, continuous, Poisson-stable function (for example, an almost periodic function). Then the operator L is regular only if it is correct or weakly regular. To prove this we must apply Theorem 2 and Lemma 2.

[1] Weak regularity (just as regularity and correctness) is preserved under small (with respect to the norm of Hom (C, C)) perturbations.

2. Suppose that the solving operator $U(t, s)$ exists for any $t, s \in J$ and satisfies the estimate

$$\|U(t, s)\| \leqslant l_1 \exp(-c_1|t-s|).$$

For any $f \in C$ the formula

$$u(t) = \operatorname{sgn} t \int_0^t U(t, s) f(s)\, ds$$

gives a bounded solution of $Lu = f$. The operator L is weakly regular but not regular. If B is one-dimensional, then an example of such an operator is $d/dt - \eta(t)$, where $\eta(t)$ is a continuous odd function such that $\eta(t) = 1$ for $t \geqslant 1$.

3. Let $U_0(t) = \exp(-tA_0)$ $(t \geqslant 0)$ be a semigroup of isometric operators. Let us assume that, for $t > 0$, $U_0(t)$ is not unitary ($B = H$ is a Hilbert space).

We prove that the operator $L = d/dt + A_0 - I$ is correct. Indeed, suppose that $Lu = f$ $(u, f \in C)$. We find a point t_0 for which $\|u(t_0)\| \geqslant \frac{1}{2}\|u\|_C$. Then from the usual formula

$$U(1)u(t_0) = u(t_0+1) - \int_{t_0}^{t_0+1} U(t_0+1-s)f(s)\, ds$$

we have $\frac{1}{2}e\|u\|_C \leqslant \|u\|_C + e\int_{t_0}^{t_0+1} \|f(s)\|\, ds$, which actually gives strong correctness. However, the operator L is non-regular. For the solutions of the homogeneous equation $Lu = 0$ increase exponentially as $t \to +\infty$. Therefore, the exponential dichotomy on J is possible if and only if these solutions can be extended to the whole line (see the definition of the exponential dichotomy on J). The latter is incompatible with the assumption that $U_0(t)$ is not unitary.

4. Let H_+ and H_- be orthogonal mutually complementary subspaces in H, and P_+ and P_- the projection operators. We consider the group of unitary operators $U_0(t)$ satisfying the conditions

$$U_0(t)H_+ \subset H_+ \quad (t \geqslant 0), \quad U_0(t)H_- \subset H_- \quad (t \leqslant 0). \tag{21}$$

In addition, we assume that H_+, H_- are not invariant under the whole group. Let A_0 be the infinitesimal generator of this group. We set $Rx = P_+x - P_-x = x_+ - x_-$, and $A = A_0 + R$. We show that (C, C) is admissible for $L = d/dt - A$, but L is not regular.

Since R is bounded, A generates a strongly continuous group $U(t)$.

From (21) we easily obtain

$$U(t)x_+ = \exp(-t)U_0(t)x_+ \quad (t \geqslant 0), \tag{22}$$

$$U(t)x_- = \exp(t)U_0(t)x_- \quad (t \leqslant 0). \tag{23}$$

Now we note that the restriction of the semigroup $U(t)$ $(t \geqslant 0)$ to H_+ is a semigroup with a generating operator A_+. In view of the exponential estimate (22), $L_+ = \mathrm{d}/\mathrm{d}t - A_+$ is regular (considered relative to H_+). Similarly, the restriction of the semigroup $U(t)$ $(t \leqslant 0)$ to H_- leads to a regular operator L_-. Now let $f \in C$ and $f = f_+ + f_-$. Then $u = L_+^{-1}f_+ + L_-^{-1}f_-$ is a bounded solution of $Lu = f$, that is, (C, C) is admissible for L.

However, the operator L is not regular. For if we were to assume regularity, then by Theorem 1 we would have the exponential dichotomy on the whole line J. It follows from (22) and (23) that the exponential dichotomy must be given by subspaces H_+ and H_-. But then these subspaces are invariant under $U(t)$ $(t \in J)$; hence we easily obtain that they are also invariant under $U_0(t)$ $(t \in J)$, which contradicts our assumption.

To realise the several conditions we can take

$$H = \mathscr{L}^2(-\infty, \infty), \quad H_+ = \mathscr{L}^2(0, \infty), \quad H_- = \mathscr{L}^2(-\infty, 0),$$

and $U_0(t)$ to be the group of right translations.

5. Instead of R in the preceding example we can take εR, where ε is small. Then from the elementary theory of perturbations of semigroups we obtain the representation $U(1) = U_0(1)(I + K)$, where K is an operator whose norm is small. Therefore, $I + K = \exp(F)$, where F is bounded. The operator $U_0(1)$ (as any unitary operator) admits the representation $U_0(1) = \exp(W)$, where W is bounded. Then, $U(1) = \exp(W)\exp(F)$. For this situation Massera & Schäffer [87] (p. 354) have indicated a constructive method of producing a periodic equation $Lu \equiv u' + A(t)u = 0$ ($A(t)$ is bounded) for which the monodromy operator is $U(1)$. L is weakly regular but not regular; this follows from the readily seen reducibility to the operator in Example 4.

6. Suppose that $A(t) = A$ is bounded, B is arbitrary, and that one of the following conditions holds: (1) $(C^+, \mathring{C})$ is admissible; (2) L is correct. We prove regularity.

For definiteness we assume that $(C^+, \mathring{C})$ is admissible. Then for each $f \in \mathring{C}$ there is a solution $u \in C^+$ such that

$$\|u\|_{C^+} \leqslant k\|f\|_C. \tag{24}$$

It suffices to prove that the spectrum of A does not intersect the imaginary axis. We assume otherwise and let $-\mathrm{i}\lambda_0$ be a boundary point of the spectrum lying on the imaginary axis. By a theorem about boundary points[2] there is a sequence $a_n \in B$ such that

$$Aa_n + \mathrm{i}\lambda_0 a_n = b_n \to 0, \quad \|a_n\| = 1. \tag{25}$$

Putting $v_n^\lambda = \exp(\mathrm{i}\lambda t)a_n$ and $\Delta = |\lambda - \lambda_0|$ (λ is real) we obtain $Lv_n^\lambda = \exp(\mathrm{i}\lambda t)(Aa_n + \mathrm{i}\lambda a_n) = f_n^\lambda$.

From (25) it follows that $\|f_n\|_C \leqslant \beta(\Delta)$, where $\lim_{\Delta\to 0}\beta(\Delta) = 0$.

Let u_n^λ be a solution of $Lu = f_n^\lambda$ satisfying the estimate (24). We set $z_n^\lambda = v_n^\lambda - u_n^\lambda$.

We define the weak topology on the dual space B^{**} of B^*; it is a fact that the unit ball in B^{**} is weakly compact. For $z \in C^+$ we denote by $m_\lambda(z)$ a weak limit point of

$$\frac{1}{T}\int_0^T z(t)\exp(-\mathrm{i}\lambda t)\,\mathrm{d}t \quad (T \to +\infty).$$

Clearly, if $Lz = 0$, then $m_\lambda(z)$ is an eigenvector of the operator A^{**} with the eigenvalue $-\mathrm{i}\lambda$; this is easily proved by taking into account the weak continuity of A^{**}. Now we note the following. Since $m_\lambda(v_n^\lambda) = a_n$ we have

$$\|m_\lambda(z_n^\lambda)\| = \|m_\lambda(v_n^\lambda) - m(u_n^\lambda)\| \geqslant 1 - k\beta(\Delta) \geqslant \tfrac{1}{2},$$

provided that $|\lambda - \lambda_0| = \Delta$ is small enough. This means that $-\mathrm{i}\lambda$ is an eigenvalue of A^{**}. Since the spectra of A^{**} and A coincide, we obtain a contradiction to the assumption that $-\mathrm{i}\lambda_0$ is a boundary point of the spectrum.

7. We study the regularity of an operator with coefficients that have small oscillations. A typical example is $\mathrm{d}/\mathrm{d}t + A(\varepsilon t)$, where ε is small (here and below $A(t)$ is assumed to be bounded).

We define a 'measure of the oscillation' of an operator-function $A(t)$ by

$$\mu(T) = \sup_{t_0\in J}\int_{-T}^{T} \|A(t_0+t) - A(t_0)\|\,\mathrm{d}t.$$

[2] See, for example, Daletskii & Krein [39], Russian p. 44, English translation, p. 28.

Lemma 3. *Let the operators* $L_{t_0} = \mathrm{d}/\mathrm{d}t + A(t_0)$ $(t_0 \in J)$ *be uniformly regular. Then if* $\mu(T)$ *is sufficiently small for a sufficiently large* T, *then* $L = \mathrm{d}/\mathrm{d}t + A(t)$ *is regular.*

Proof. From (17) it follows that the norms of the operators $L_{t_0}^{-1} : M^1 \to C$ are not greater than some constant k_1.

Let $Lu = f$, where $f \in M^1$ and $u \in C$. We take a point $t_0 \in J$ for which $\|u(t_0)\| \geq \frac{1}{2}\|u\|_C$, and rewrite the equation $Lu = f$ in the form $L_{t_0}u = (A(t_0) - A(t))u + f$. We selected a ϕ for which $\phi(t_0) = 1$, supp $\phi \in [-T + t_0, T + t_0]$, and $|\phi'(t)| \leq 2/T$. For $v = \phi u$ we have

$$L_{t_0}v = \phi(t)(A(t_0) - A(t))u + \phi f + \phi' u = g(t).$$

Therefore,

$$\begin{aligned}\tfrac{1}{2}\|u\|_C \leq \|v\|_C \leq k_1\|g\|_{M^1} \leq k_1\mu(T)\|u\|_C + k\|f\|_{M^1} \\ + k_1(2/T)\|u\|_{M^1}.\end{aligned}$$

Assuming that $1/T$ and $\mu(T)$ are sufficiently small, we hence obtain that $\|u\|_C \leq \frac{1}{4}k\|f\|_{M^1}$, which gives, in particular, the strong correctness of L.

Because these arguments can be applied to the dual operator L^*, by Theorem 3 the operator L is regular.

Corollary 1. *Suppose that* $A(t)$ *is a uniformly continuous function in* Hom (B, B) *and the operators* L_{t_0} *are uniformly regular. Then the operator* $\mathrm{d}/\mathrm{d}t + A(\varepsilon t)$ *is regular for small* ε.

Corollary 2. *Suppose that the function* $A(t)$ $(t \geq 0)$ *is compact in* Hom (B, B) *and 'stationary at infinity', that is,*

$$\lim_{t_0 \to \infty} \int_{-T}^{T} \|A(t + t_0) - A(t_0)\| \,\mathrm{d}t = 0 \quad (T > 0).$$

Suppose that the spectrum of each of the 'limiting' operators $\hat{A}$ $(\hat{A} = \lim_{t_n \to \infty} A(t_n))$ does not intersect the imaginary axis. Then the exponential dichotomy on $J^+ = [0, +\infty)$ holds for $L = \mathrm{d}/\mathrm{d}t + A(t)$.

For a proof we must choose $t_0 > 0$ large enough that the operators L_s $(s \geq t_0)$ are uniformly regular (we can do this in view of the compactness) and that the operator-function

$$A_1(t) = \begin{cases} A(t) & \text{for } t \geq t_0, \\ A(t_0) & \text{for } t \leq t_0, \end{cases}$$

satisfies the conditions of Lemma 3. Then the operator L_1 has the exponential dichotomy on J (by Theorem 1), and consequently, L has it on J^+.

Comments and references to the literature

Questions of admissibility and dichotomy closely connected with the general theory of stability are treated in the books of Massera & Schäffer [87], Daletskii & Krein [39], and Krasnosel'skii, Burd & Kolesov [66]. The first two of these contain extensive bibliographies and historical notes.

§ **2.** A connection between regularity and the exponential dichotomy in the finite-dimensional case was essentially known to Perron [98] (Theorem 1 is explicitly in the article of Maizel' [84]). But there was no similar result for equations in a Banach space. Massera & Schäffer deduced the exponential dichotomy from regularity by assuming a certain 'closedness condition' (regarding this, see also Daletskii & Krein [39], Russian p. 249, English tr. p. 203). The observation that 'conditions of closedness' are redundant was made by Zhikov [55].

§ **3.** We mention the interesting article of Mukhamadiev [94] who studied a property fairly close to correctness. We consider the operator $L = \mathrm{d}/\mathrm{d}t + A(t)$, where $A(t)$ is an almost periodic matrix-function, and assume that equations of the form $Lu \equiv u' + \hat{A}(t)u = 0$ do not have bounded solutions. Then L is regular. This result of Mukhamadiev can be reduced to Theorem 2 by proving that L is correct. Indeed, if L is not correct, then there are $u_n, f_n \in C$ such that

$$Lu_n = f_n, \quad \|u_n\|_C = 1, \quad \|f_n\|_C \to 0.$$

We choose points $t_n \in J$ for which $\|u_n(t_n)\| \geqslant \frac{1}{2}$. By taking the limit in $\{u_n(t + t_n)\}$ we obtain a non-trivial bounded solution of one of the limiting equations.

§ **4.** The results in the example in § 4.6 were inspired by a question posed by Daletskii & Krein [39] (Russian p. 44, English translation p. 27). The results in § 4.7 were obtained jointly by Zhikov and Valikov, while the other results in § 4 are due to Zhikov [55].

The reader can find further discussions in the article of Zhikov & Tyurin [58].

11 The averaging principle on the whole line for parabolic equations

1 Bogolyubov's lemma

We consider the linear operators

$$L_\omega = \mathrm{d}/\mathrm{d}t + A(\omega t),$$

where $A(t)$ is a bounded continuous function $J \to \mathrm{Hom}\,(X, X)$, X being some Banach space. We are interested in the properties of the operator L_ω as $\omega \to \infty$. Suppose that $A(t)$ is such that the mean

$$\lim_{T\to\infty} \frac{1}{2T} \int_{-T+\alpha}^{T+\alpha} A(t)\,\mathrm{d}t = \bar{A} \quad \text{(uniformly with respect to } \alpha \in J)$$

exists in the sense of the operator norm. We introduce the so-called *averaged operator* $\bar{L} = \mathrm{d}/\mathrm{d}t + \bar{A}$.

Bogolyubov's lemma. *If the averaged operator $\bar{L}$ is regular, then the operators L_ω are uniformly regular for $\omega \geqslant \omega_0$.*

Proof. We put $\tilde{A}(t) = A(t) - \bar{A}$. The function $A(t)$ has a zero uniform mean; by Lemma 2 and Theorem 3 in Chapter 6 there is a family of functions $\tilde{A}_\varepsilon(t) : J \to \mathrm{Hom}\,(X, X)$ such that

$$\lim_{\varepsilon\to\infty} \sup_{t\in J} \|\tilde{A}(t) - \tilde{A}_\varepsilon(t)\| = 0, \quad \sup_{t\in J} \left\| \int_0^t \tilde{A}_\varepsilon(s)\,\mathrm{d}s \right\| < \infty.$$

We put $B_\varepsilon(t) = \int_0^t \tilde{A}_\varepsilon(s)\,\mathrm{d}s$; and consider the operator $T : C(X) \to C(X)$ defined by

$$x(t) \to x(t) - \omega^{-1} B_\varepsilon(t\omega) x(t).$$

We have

$$\begin{aligned} L_\omega T &= (\mathrm{d}/\mathrm{d}t + \bar{A} + \tilde{A}(t\omega))(I - \omega^{-1} B_\varepsilon(t\omega)) \\ &= (I - \omega^{-1} B_\varepsilon(t\omega))\,\mathrm{d}/\mathrm{d}t + \bar{A} + (\tilde{A}(t\omega) - \tilde{A}_\varepsilon(t\omega)) \\ &\quad - \omega^{-1} A(\omega t) B_\varepsilon(t\omega). \end{aligned}$$

We can make the third term on the right-hand side small by a suitable choice of ε, and the fourth by a choice of ω. As a result we obtain that the operator $T^{-1}L_\omega T$ is represented in the form $\bar{L}+\delta$, where δ is a small perturbation. Since regularity is preserved under small perturbations, $T^{-1}L_\omega T$, and hence L_ω, is regular for large ω. This proves the lemma.

In Bogolyubov's lemma we have used the boundedness of the operators $A(t)$. We show by an example that boundedness is really essential; in this example we have:

1°. The unbounded operator $\bar{A}$ generates an exponentially stable group, that is, $\bar{L}=\mathrm{d}/\mathrm{d}t+\bar{A}$ is automatically regular.

2°. $A(t)-\bar{A}$ is a continuous periodic function $J \to \operatorname{Hom}(X, X)$ with zero mean.

3°. The operator $L_\omega=\mathrm{d}/\mathrm{d}t+A(\omega t)$ is not regular as $\omega \to \infty$.

Example. In a Hilbert space H_α (α is a real parameter) we determine an orthonormal basis $\{e_n^\alpha\}$ $(n=0, 1, \ldots)$. Linear operators $\bar{A}$ and K are defined by:

$$\bar{A}e_n^\alpha=(\mathrm{i}\alpha n-1)e_n^\alpha,\quad Ke_n^\alpha=e_{n+1}^\alpha.$$

As a preliminary let us note that the operator

$$L=\mathrm{d}/\mathrm{d}t+\bar{A}+\exp(-\mathrm{i}\alpha t)K$$

is not invertible in $C=C(H_\alpha)$. For let $f(t)=-e_0$. If $Lu=f$ has a solution $u\in C$, then by writing $u(t)=\sum_{n=0}^{\infty}u_n(t)e_n^\alpha$ we have

$$u_0'-u_0=-1,\quad u_1'+(\mathrm{i}\alpha-1)u_1=-\exp(-\mathrm{i}\alpha t)u_0$$

$$u_n'+(\mathrm{i}n\alpha-1)u_n=-\exp(-\mathrm{i}\alpha t)u_{n-1}\quad (n=1, 2, \ldots).$$

Hence we obtain $u_0=1$, $u_1=\exp(-\mathrm{i}\alpha t), \ldots, u_n=\exp(-\mathrm{i}\alpha nt)$, that is, the equation $Lu=f$ does not have solutions in C.

Then for X we take the direct sum of a countable number of spaces H_{α_m}, where $\alpha_m \to +\infty$. We define the operators $\bar{A}$ and K in each H_{α_m} by the above formulae. Then we find that the operator $\mathrm{d}/\mathrm{d}t+\bar{A}+\exp(\mathrm{i}\omega t)K$ is not invertible in $C=C(X)$, at least for the values $\omega=\alpha_m$. Nevertheless, the averaged operator $\bar{L}$ is regular.

2 Some properties of parabolic operators

1. Let E be a reflexive embedded space; this means that there is a Hilbert space H such that the embedding $E\subset H\subset E^*$ is dense and continuous, and the bilinear form (y, x) $(y\in E^*, x\in E)$ coincides with the scalar product on H for $x, y\in H$. We shall use the following notation for norms: $\|x\|$ is an E-norm, $\|y\|_*$ is an E^*-norm, and $|x|$ is an H-norm.

A bounded linear operator $A:E \to E^*$ is called *coercive* (or *strongly elliptic*) if

$$\operatorname{Re}(Ax, x) \geqslant c_1\|x\|^2 - c_2|x|^2 \quad (c_1, c_2 > 0, x \in E).$$[1]

It is obvious that the adjoint operator $A^*: E \to E^*$ is also strongly elliptic. We can regard A as being an unbounded operator in H if we define its domain by $\{x \in H: Ax \in H\}$. This will always be understood when we speak about the spectrum of A.

We call an operator $\mathrm{d}/\mathrm{d}t + A(t)$ *parabolic* if $A(t)$ is a continuous bounded function in $\operatorname{Hom}(E, E^*)$ and the coercive inequality holds uniformly with respect to $t \in J$. We characterise a parabolic operator by the constants c_1 and c_2 in the coercive inequality and by the constant $l = \sup_{t \in J} \|A(t)\|_{E \to E^*}$.

The solvability of the Cauchy problem for a parabolic operator is readily obtained from the results of Chapter 9. For consider the operator

$$\exp(-\lambda t)I(\mathrm{d}/\mathrm{d}t + A(t)) \exp(\lambda t)I = (\mathrm{d}/\mathrm{d}t + A(t) + \lambda I).$$

If $\lambda \geqslant c_2$, then the operators $A(t) + \lambda I$ are monotonic. Therefore, Theorem 5 of Chapter 9 with $p = q = 2$ holds. But then Theorem 5 holds for the original operator L.

Now we write out the estimates we shall need later. By scalar multiplying the equation $u' + A(t)u = f$ by $u(t)$ and integrating we obtain (by analogy with (11) in Chapter 9)

$$|u(t)|^2 \leqslant |u(t_0)|^2 + c_1^*\left(\int_{t_0}^{t} \|u\|^2\, \mathrm{d}s + \int_{t_0}^{t} \|f\|_*^{\,2}\, \mathrm{d}s\right) \quad (t \geqslant t_0), \tag{1}$$

$$\int_{t_0}^{t_0+1} (\|u\|^2 + \|u'\|_*^{\,2})\, \mathrm{d}s \leqslant c_1^*\left(|u(t_0)|^2 + \int_{t_0}^{t_0+1} \|f\|_*^{\,2}\, \mathrm{d}s\right). \tag{2}$$

Here the constant c_1^* depends only on the constants c_1, c_2 and l. Let us note that if we introduce a parameter ω in the operator A, then it turns out that (1) and (2) are uniform with respect to ω.

[1] It is worth emphasising that in applications, usually beforehand we have only the embedding $E \subset H$ and a bilinear form $a(u, v)$ continuous on E. Then E^* is determined as the closure of H in the norm

$$\|y\|_* = \sup_{\|x\| \leqslant 1} |(y, x)|,$$

and the operator $A: E \to E^*$ is defined by $(Au, v) = a(u, v)$.

2. The operator $L^* = -\mathrm{d}/\mathrm{d}t + A^*(t)$ is by virtue of its properties, analogous to L, except that the Cauchy problem is solvable from the left. We show that L and L^* are duals, that is, they are connected by Green's formula.

Let $Lu = f$ and $L^*v = g$ $(a \leq t \leq b)$, where $f, g \in \mathcal{L}^2(a, b; E^*)$. From results in § 2 in Chapter 9 it follows that

$$\mathrm{d}(u, v)/\mathrm{d}t = (u', v) + (u, v')$$

almost everywhere on [a, b]. Therefore,

$$\begin{aligned}\mathrm{d}(u, v)/\mathrm{d}t &= (-Au + f, v) + (u, A^*v - g)\\ &= (f, v) - (u, g);\end{aligned}$$

$$(u, v)|_a^b = \int_a^b \{(Lu, v) - (L^*v, u)\}\, \mathrm{d}t,$$

as we required.

We define the concepts of regularity, correctness, and weak regularity relative to the space $C = C(H)$.

Corollary 1. *Suppose that the embedding $E \subset H$ is compact and $A(t)$ is a Poisson-stable function in* Hom (E, E^*) *(for example, an almost periodic function). Then the operator L is regular whenever it is correct or weakly regular.*

Proof. We apply Theorem 2 of Chapter 10. Since the compactness condition follows from the compactness of the embedding $E \subset H$, we must verify the P-property.

Let $t_m \to -\infty$ and $A(t + t_m) \to_{\mathrm{loc}} A(t)$ $(t \in J)$. We set $L_m = \mathrm{d}/\mathrm{d}t + A(t + t_m)$. Suppose that $u(t)$ and $u_m(t)$ $(t \geq 0)$ are solutions of $Lu = 0$ and $L_m u = 0$ with a common initial condition $u_m(0) = u(0) = 0$. We must prove that $|u(t) - u_m(t)| \to 0$ for $t \geq 0$.

From the relation $L(u - u_m) = (A(t) - A(t + t_m))u_m$ and the estimates (1) and (2), for $0 \leq t \leq T$ we have

$$|u(t) - u_m(t)|^2 \leq c_1{}^* \sup_{0 \leq t \leq T} \|A(t + t_m) - A(t)\|_{E \to E^*} \int_0^T \|u_m\|^2\, \mathrm{d}s,$$

which proves the P-property. Weak regularity is dealt with by using Lemma 2 of Chapter 10.

3. In what follows the space E^* is subject to the following condition: there is a sequence of linear operators $P_m : E^* \to E^*$ such that

$$P_m E^* \subset E, \quad P_m y \xrightarrow{E^*} y \quad (y \in E^*). \tag{3}$$

We introduce the following spaces of functions: $X = M^2(E^*)$, X_{comp} is the subspace of X consisting of those $f(s) \in X$ for which the family of translates $f(t+s)$ $(t \in J, 0 \leq s \leq 1)$ is compact in $\mathscr{L}^2(0, 1; E^*)$; $V = M^2(E) \cap C(H)$ with the norm

$$\|u\|_V = \sup_{t \in J} \left(\int_0^1 \|u(t+s)\|^2 \, ds \right)^{1/2} + \sup_{t \in J} |u(t)|;$$

W is the space of those $u \in V$ for which $u' \in X$ with the norm given by

$$\|u\|_W = \|u\|_V + \|u'\|_X.$$

In these spaces the norm is translation-invariant. This enables us in an intrinsic way to distinguish subspaces $\mathring{X}$, $\mathring{V}$, $\mathring{W}$ consisting of almost periodic elements. For instance, $f \in X$ is almost periodic if $\{f(t+s)\}$ is compact in X.

In terms of the above spaces, the estimates (1) and (2) have the form

$$\|u\|_V^2 \leq \|u\|_W^2 \leq c_1{}^* \{\|u\|_{C(H)}^2 + \|f\|_X^2\}. \tag{4}$$

3 The linear problem about averaging

1. We assume that the embedding $E \subset H$ is compact, and that $A(t)$ is a compact function in Hom (E, E^*) with a uniform mean $\bar{A}$.

Theorem 1. *Suppose that the averaged operator $\bar{L}$ is regular, that is, the spectrum of A does not intersect the imaginary axis. Then the operators L_ω $(\omega \geq \omega_0)$ are uniformly regular. Additionally, the following assertions hold:*

(1) *The operator L_ω $(\omega \geq \omega_0)$ realises a homeomorphism $W \to X$, and the inverses are uniformly bounded; in particular*

$$\|u\|_V \leq k\|L_\omega u\|_X. \tag{5}$$

(2) *If $f \in X_{\text{comp}}$ and $\bar{f}$ is the uniform mean in E^*, then*

$$\lim_{\omega \to \infty} \|L_\omega{}^{-1} f(\omega t) - \bar{A}^{-1} \bar{f}\|_V = 0. \tag{6}$$

Proof. (1) Let $C = C(H)$. First we prove the estimate

$$\|u\|_C \leq k\|L_\omega u\|_X \quad (u \in C, Lu \in X, \omega \geq \omega_0). \tag{7}$$

Assuming that this estimate does not hold, we find v_n, g_n and ω_n such that

$$L_{\omega_n} v_n = g_n, \quad \|v_n\|_C = 2, \quad \|g_n\|_X \to 0 \quad \text{as } \omega_n \to \infty.$$

We choose a $t_n \in J$ for which $|v_n(t_n)| \geq 1$ and put

$$u_n(t) = v_n(t+t_n), \quad \tilde{A}_n(t) = A(\omega_n(t+t_n)),$$
$$f_n(t) = g_n(t+t_n).$$

Then $u_n' + Au_n + \tilde{A}_n u = f_n$.

The sequence $\{u_n\}$ is bounded in W (the estimate (4)). Therefore, $\{u_n\}$ and $\{u_n'\}$ are compact in the sense of weak convergence in $\mathscr{L}^2(-T, T; E)$ and $\mathscr{L}^2(-T, T; E^*)$, respectively. By going over to subsequences if necessary, we can assume the weak convergence $u_n \to z$, $u_n' \to z'$. We show that $\bar{L}z = 0$ $(t \in J)$. To this end we consider the identity

$$\int_J (u_n', \xi)\,dt + \int_J (\bar{A}u_n, \xi)\,dt = -\int_J (\tilde{A}_n u_n, \xi)\,dt + \int_J (f_n, \xi)\,dt, \tag{8}$$

where $\xi(t)$ is a fixed smooth function with compact support.

We denote the first term on the right by λ_n and prove that $\lambda_n \to 0$ (it is obvious that the second term tends to zero).

We compare $\lambda_n = \int_J (u_n, \tilde{A}_n{}^*\xi)\,dt$ with

$$\lambda_n^m = \int_J (u_n, P_m \tilde{A}_n{}^*\xi)\,dt. \tag{9}$$

Since the convergence $P_m y \to y$ (see (3)) is uniform on compact sets in E^*, and since $A^*(t)$ is a compact function in Hom (E, E^*), the limit

$$\lim_{m\to\infty} \|\tilde{A}_n{}^*\xi - P_m \tilde{A}_n{}^*\xi\|_* = 0$$

exists uniformly in n and t. Hence (and from the boundedness of u_n in V) it follows that $\lim_{m\to\infty} \lambda_n^m = \lambda_n$ uniformly with respect to n. Therefore, it is enough to prove that $\lim_{n\to\infty} \lambda_n^m = 0$ (m is fixed).

We set $B_n(t) = \int_0^t \tilde{A}_n(s)\,ds$. Since

$$B_n(t) = \int_0^t (A(\omega_n(s+t_n)) - \bar{A})\,ds = \frac{1}{\omega_n}\int_{t_n\omega_n}^{t_n\omega_n + t\omega_n} (A(s) - \bar{A})\,ds,$$

from the uniformity of the mean $\bar{A}$ we obtain

$$\|B_n(t)\|_{E\to E^*} \xrightarrow{\text{loc}} 0.$$

By integrating (9) by parts we obtain

$$\lambda_n^m = -\int_J (u_n', P_m B_n{}^*\xi)\,dt - \int_J (u_n, P_m B_n{}^*\xi')\,dt.$$

Since $\|P_m B_n{}^*(t)\|_{E\to E} \xrightarrow{}_{\text{loc}} 0$, it is now obvious that $\lambda_n^m \to 0$ as $n \to \infty$. Thus we can take the limit in the identity (8); as a result we obtain $\bar{L}z = 0\,(t \in J)$. We show that $z(t) \not\equiv 0$. Since $\|z\|_C \leqslant 2$, this contradicts the regularity of $\bar{L}$.

By Lemma 3 of Chapter 9, the sequence $\{u_n\}$ is compact in $\mathscr{L}^2(-T, T; H)$. Since we already have weak convergence, $u_n(t) \to z(t)$ in $\mathscr{L}^2(-T, T; H)$. Recall that $|u_n(0)| \geqslant 1$. Allowing that $z(t) \equiv 0$, from (1) we have

$$|u_n(0)|^2 \leqslant |u_n(t)|^2 + c_1 \int_{-T}^{0} (|u_n|^2 + \|f_n\|_*^2)\, ds \quad (-T \leqslant t \leqslant 0).$$

The integral on the right-hand side tends to zero (by assumption), therefore $|u_n(t)| \geqslant \frac{1}{2}$ provided that n is large enough; but then $|u_n(t)\| \to 0$ in $\mathscr{L}^2(-T, 0)$ is impossible. This proves (7).

2. The estimate (7) means, in particular, that the operators L_ω are strongly correct, since there is a natural embedding $M^2(H) \subset M^2(E^*) = X$. To prove regularity we use Theorem 3 of Chapter 10. For this it is enough to note that the dual operator L^* is regular together with $\bar{L}$, and that the preceding arguments could be repeated for the operators $L_\omega{}^*$, that is, the $L_\omega{}^*(\omega \geqslant \omega_0)$ are strongly correct. Then the operators $L_\omega(\omega \geqslant \omega_0)$ are regular.

3. Now we regard L_ω as an operator $W \to X$. From the estimates (7) and (4) we obtain that

$$\|u\|_W \leqslant k_1 \|L_\omega u\|_X. \tag{10}$$

In Chapter 10 we proved assertions of the following type: the range of a correct operator is dense in C, and the range of a strongly correct operator, in M^2. Similarly it can be proved that the range of an operator with the estimate (10) is locally closed in X. Since the range of L_ω already contains $C(H)$ (regularity), it must coincide with X. This proves assertion (1).

4. The operator $\bar{A} : E \to E^*$ is a homeomorphism. In fact, the preceding arguments could be carried out not for L_ω but for $\bar{L}$, that is, $\bar{L} : W \to X$ is a homeomorphism. In particular, if $f \equiv y \in E^*$, then it is clear that $x = \bar{L}^{-1} f$ does not depend on t, that is, $\bar{A}x = y$, $x \in E$. Now we prove assertion (2). We put

$$u^\omega = L_\omega{}^{-1}(f(\omega t)) - \bar{A}^{-1}\bar{f}.$$

Obviously, $L_\omega u = f(\omega t) - \bar{f} = g(\omega t)$.

Let $\hat{X}_{\text{comp}}$ stand for the subspace of elements of X_{comp} having a uniform mean in E^*. Since the estimate (5) is already proved, it is sufficient to prove (6) for a set dense in $\hat{X}_{\text{comp}}$; we obtain such a dense set in the following way. For any $f \in \hat{X}_{\text{comp}}$ we consider the relation $P_m f(t+s) \to f(t+s)$. Since we have convergence in $\mathscr{L}^2(0, 1; E^*)$ for each $t \in J$ (see property (3)), in view of compactness this convergence is uniform with respect to t. Thus, f is approximated by an element in $\hat{M}_{\text{comp}}(E)$. Then by using Steklov averaging in a similar sense we obtain an approximation by an element in $C(E)$ with a uniform mean. Furthermore, by Lemma 2 and Theorem 3 of Chapter 6, we have a best possible approximation by an element of the form $\bar{f} + g(t)$, where $\bar{f} \in E$, $g(t) \in C(E)$ and $\eta(t) = \int_0^t g(s)\, ds \in C(E)$.

After this we make the substitution $u = z + \omega^{-1}\eta(\omega t)$ in the equation $L_\omega u^\omega = g(\omega t)$ to give

$$L_\omega z^\omega = \omega^{-1} A(\omega t) g(\omega t).$$

But then $\|z^\omega\|_V \to 0$, and so from (5) it follows that $\|u^\omega\|_V \to 0$. Theorem 1 is completely proved.

Remark **1.** If $A(t)$ is an almost periodic function in Hom (E, E^*), then in Theorem 1 we can take $\mathring{X}$, $\mathring{V}$, $\mathring{W}$ instead of X, V, W. The proof is straightforward and we omit it.

Remark **2.** We return to Bogolyubov's lemma and discuss when similar arguments can be applied to a parabolic operator.

As we have already noted, an averaged regular operator $\bar{L}$ realises a homeomorphism $W \to X$, which is obviously preserved under perturbations that are small with respect to the norm of Hom (W, X). Let us assume that $\tilde{A}(t) = A(t) - \bar{A}$ is a completely continuous operator $E \to E^*$ for every $t \in J$. Then $\tilde{A}(t)$ can be represented as a sum of a 'small' perturbation and a finite-dimensional operator-function. This enables us to repeat the proof of Bogolyubov's lemma and to find a transformation T for which $L_\omega T = \bar{L} + \delta$, where δ is a small perturbation. It can be shown that the converse also holds: if there exists a transformation T with the indicated property, then $A(t) - \bar{A}$ is a completely continuous operator $E \to E^*$.

4 A non-linear equation

By using Theorem 1 the construction of the non-linear theory is completely trivial; the only point requiring some attention is that in studying the properties of conditional stability it is not possible to use the usual apparatus of differential inequalities. We resolve this minor deficiency by some very simple additional analysis of a linear approximation in special spaces.

We introduce the spaces X_λ, V_λ $(\lambda \geq 0)$ consisting of those $f \in X$, $u \in V$ for which the following are finite:

$$\|\exp(\lambda|t|)f(t)\|_X, \quad \|\exp(\lambda|t|)u(t)\|_V;$$

these are used as norms in X_λ, V_λ, respectively. We also introduce the spaces X^+, V^+, X_λ^+, V_λ^+, which consist of functions on J^+ and are the analogues of X, V, X_λ, V_λ.

Suppose that the parabolic operator L has an inverse $L^{-1}: X \to V$; we denote its norm by k_0.

Lemma 1. *There is a* $\lambda_1 > 0$ *such that for* $\lambda \in [0, \lambda_1]$ *the operators* $L^{-1}: X_\lambda \to V_\lambda$ *exist and their norms are not greater than* $2k_0$.

Proof. Let $f \in X_\lambda$ and $u = L^{-1}f$. We put

$$\phi_n(t) = \begin{cases} \exp(\lambda|t|) & \text{for } |t| \leq n, \\ \exp(\lambda n) & \text{for } |t| \geq n. \end{cases}$$

Then $v_n = \phi_n u$ satisfies $Lv_n = \phi_n f + \phi_n' u$. Therefore,

$$\|v_n\|_V \leq k_0\{\|\phi_n f\|_X + \|\phi_n' u\|_X\} \leq k_0\{\|\phi_n f\|_X + \lambda\|\phi_n u\|_X\}.$$

Since $\|\phi_n u\|_X \leq \|\phi_n u\|_V$, for $k\lambda \leq \frac{1}{2}$ we obtain $\|\phi_n u\|_V \leq \frac{1}{2}k_0\|\phi_n f\|_X$. The required inequality is obtained by taking the limit, and Lemma 1 is proved.

Next, under the conditions of Theorem 1 the operators L_ω have the uniform exponential dichotomy on J (see Chapter 10, § 2). Let $N^{\omega_1}(t_0)$ and $N^{\omega_2}(t_0)$ be the subspaces giving this dichotomy, $P^{\omega_1}(t_0)$ and $P^{\omega_2}(t_0)$ be the corresponding projection operators, and c_1 the constant in conditions (d_3) and (d_4) in Chapter 10. We put $\lambda_0 = \min\{\lambda_1, c_1\}$, and extend an arbitrary $f \in X_{\lambda_0}^+$ by zero to the whole line; let $u^\omega = L_\omega^{-1} f$. We set by definition

$$T^\omega{}_+ f = u^\omega(t) - U^\omega(t,0)P^\omega{}_1(0)u^\omega(0) \quad (t \geq 0).$$

The fact that we obtain a bounded family of operators $T^\omega{}_+ : X_{\lambda_0}^+ \to V_{\lambda_0}^+$ follows from Lemma 1 and the inequality

$$\|U^\omega(t,0)a\|_{V_{\lambda_0}{}^+} \leq l|a| \quad (a \in N^\omega{}_1(0)), \tag{11}$$

which in turn is easily derived from condition (d_3) and the *a priori* estimate (2).

Suppose that the norms of L_ω^{-1} and $T^\omega{}_+$ are not greater than a number k. We consider the non-linear equation

$$u' + F(u, \omega t) = 0. \tag{12}$$

Let there be a point $\mathring{x} \in E$ for which $f(t) = -F(\mathring{x}, t)$ is an element of X_{comp} with zero uniform mean, and assume that the following conditions hold:

(1) The operators $F(x, t): E \to E^*$ are Fréchet differentiable at $\mathring{x}$.

(2) The operator $L = \mathrm{d}/\mathrm{d}t + A(t)$, where $A(t) = F_x'(\mathring{x}, t)$, satisfies the conditions of Theorem 1.

(3) The operator $F(u(t), t)$ acts from V into X, and is continuously differentiable in some V-neighbourhood of $\mathring{x}$.

Remark. Since there are natural embeddings $E \subset V$ and $E^* \subset X$, then condition (1) follows from condition (3), at least when $F(x, t) \equiv F(x)$ (the converse is not true).

Under conditions (1)–(3) we have

Theorem 2. *For $\omega \geqslant \omega_0$ equation (12) has a unique solution $u^\omega \in V$ such that $\|u^\omega - \mathring{x}\|_V \to 0$. This solution is conditionally exponentially stable uniformly with respect to the initial value $t_0 \in J$ and to $\omega \geqslant \omega_0$. In particular, for any $a \in N^{\omega_1}(t_0)$ there is a unique solution $y^\omega(t)$ $(t \geqslant t_0)$ for which $P^{\omega_1}(t_0)\,(y^\omega(t_0) - u^\omega(t_0)) = a$, and*

$$\|y^\omega - u^\omega\|_{V_\beta^+} \leqslant l_0 |a|$$

whenever $|a| \leqslant \rho_0$. The constants l_0 and ρ_0 do not depend on t_0 and ω.[2]

Proof. For convenience we take $\mathring{x} = 0$; we rewrite (12) as

$$L_\omega u = -F(u, \omega t) + A(\omega t)u = \Phi^\omega(u).$$

The operator Φ^ω satisfies the Lipschitz condition

$$\|\Phi^\omega(u_1) - \Phi^\omega(u_2)\|_{X_\lambda} \leqslant (1/2k)\|u_1 - u_2\|_{V_\lambda} \quad (\lambda \geqslant 0), \tag{13}$$

whenever $\|u_1\|_V$ and $\|u_2\|_V \leqslant r_0$. For $\lambda = 0$ this follows from the identity

$$\Phi^\omega(u_1) - \Phi^\omega(u_2) = \int_0^1 \Phi_u^\omega(u_2 + s(u_1 - u_2))(u_1 - u_2)\,\mathrm{d}s,$$

since $\|\Phi_u^\omega(v)\|_{\mathrm{Hom}(V,X)} \to 0$ as $\|v\|_V \to 0$ (condition (3)). For $\lambda > 0$ we need to multiply this identity by $\exp(\lambda|t|)$ and use the fact that the derivative Φ_u^ω commutes with multiplication by a scalar function. We prove the existence of a solution $u^\omega \in V$. To this end, we consider in the space V the equation

$$u = L_\omega^{-1}\{\Phi^\omega(u)\} = Q^\omega(u).$$

For the operator Q^ω the Lipschitz constant is $\frac{1}{2}$ provided that $\|u\|_V \leqslant r_0$.

[2] We have stated only part of the property of conditional stability; the rest is formulated and proved similarly.

Since the first approximation $u^{\omega_1}=L_\omega^{-1}\{f(\omega t)\}$ satisfies the condition $\|u^{\omega_1}\|_V \to 0$ (see (6)), we obtain the required solution by iterations.

Now we establish conditional stability, assuming for convenience that $t_0=0$. To prove the existence of solutions $y^\omega(t)$ $(t\geqslant 0)$ we consider the equation $y=U^\omega(t,0)a+T^\omega{}_+\{\Phi^\omega(y)\}$. By making the substitution $z=y-u^\omega$ we obtain

$$z=U^\omega(t,0)a+T^\omega{}_+\{\Phi^\omega(u^\omega+z)-\Phi^\omega(u^\omega)\}=D^\omega(z).$$

Let us estimate the Lipschitz constant of D^ω in $V_{\lambda_0}{}^+$. Since $\|u^\omega\|_V\to 0$, it follows from (13) that this constant is not greater than $\frac{1}{2}$ provided $\|z\|_{V_{\lambda_0}+}\leqslant r_1$. Therefore the iteration process converges if the first approximation $z^{\omega_1}=D^\omega(z^{\omega_0})=D^\omega(0)=U^\omega(t,0)a$ is sufficiently small in V_{λ_0}, that is, if $|a|\leqslant\rho_0$ (see (11)). Here the fixed point (as is always the case in the contraction mapping principle) satisfies the estimate

$$\|z\|_{V_{\lambda_0}+}=\|z^\omega-z^{\omega_0}\|_{V_{\lambda_0}+}\leqslant 2\|z^{\omega_1}-z^{\omega_0}\|_{V_{\lambda_0}+}\leqslant 2\|z^{\omega_1}\|_{V_{\lambda_0}+}\leqslant l_0|a|.$$

This proves Theorem 2.

Unfortunately, the actual range of application of Theorem 4 is quite narrow; the situation is that the differentiability condition (condition 3) is delicate. It is not difficult to give the cases (relative to non-linear parabolic equations) when this condition automatically does not hold. For this it is useful to bear in mind the following observations.

1°. The operator $F:B_1\to B_2$ (assumed to be independent of t and such that $F(0)=0$) generates a differentiable mapping $\mathscr{L}^2(0,1;B_1)\to\mathscr{L}^2(0,1;B_2)$ only when it is linear. Hence we obtain the following. Suppose that $F:E\to E^*$ is a non-linear operator. Then we can establish its differentiability as an operator $V\to X$ only at the expense of requiring that the term $\|u\|_C$ occurs in the expression for $\|u\|_V$.

2°. The superposition operator $G(\phi(x))$ gives a differentiable mapping $\mathscr{L}^2(\Omega)\to\mathscr{L}^2(\Omega)$ only when G is a linear function.[3]

We denote by $H_s(\Omega)$ $(s=0,1,\ldots)$ the Sobolev space of functions $u(x)$ with the norm

$$\int_\Omega \sum_{j\leqslant s}|u^{(j)}(x)|^2\,\mathrm{d}x,$$

where $u^{(j)}$ denotes the derivative of order j and the summation is over all derivatives. We consider the Dirichlet problem for a second-

[3] Here and in what follows Ω is a finite domain in the Euclidean space of variables $x=(x_1,\ldots,x_m)$.

order equation. Then $H=H_0=\mathscr{L}^2(\Omega)$, $E=\mathring{H}_1$ ($\mathring{H}_1$ is the closure in H_1 of the set of infinitely smooth functions with compact supports), and typical non-linearities are given by expressions of the form

$$\frac{\partial}{\partial x}G(u_x),\quad G(u_x),\quad \frac{\partial G}{\partial x}(u),\quad G(u).$$

We consider the differentiability of these operators. The first cannot be differentiable (not being linear) even as an operator $E\to E^*$; this follows easily from 2°. The second can define a differentiable mapping $E\to E^*$, but not $V\to X$. A rigorous proof is fairly lengthy, but it is easy to construct examples of non-differentiable operators with an infinitely smooth G with compact support. There remain the operators $\partial G(u)/\partial x$ and $G(u)$. We discuss the operator $\partial G(u)/\partial x$ in detail. Suppose that $|G'(u)|\leqslant l\{1+|u|^\lambda\}$ ($\lambda\geqslant 0$). Then it can be proved that:

3°. The operator $\partial G(u)/\partial x : E\to E^*$ exists and is continuously differentiable for $m\geqslant 3$ if $\lambda\leqslant 2/(m-2)$, and for $m=1, 2$ if $\lambda<\infty$; moreover the derivative is completely continuous.

4°. The operator $\partial G(u)/\partial x : V\to X$ exists and is continuously differentiable if $\lambda\leqslant(3-m)/(m-1)$.

5 The Navier–Stokes equation

Consider the Navier–Stokes equation

$$\begin{gathered}\frac{\partial u_j}{\partial t}-\mu\Delta u_j+\frac{\partial}{\partial x_j}(u_iu_j)=-\frac{\partial p}{\partial x_j}(x,\omega t)+f(x,\omega t),\\ \operatorname{div} u=0,\quad u|_{\partial\Omega}=0,\quad i,j=1,2,\dots,m.\end{gathered}\tag{14}$$

We denote by $\mathscr{L}^2(\Omega)$ and $H_s(\Omega)$ the spaces of vector-functions $u=(u_1,\dots,u_m)$ with components from the corresponding spaces of scalar functions. We are only interested in the cases $m=2$ and 3. We consider the set of all smooth solenoidal vectors with compact support and denote its closure in the $\mathscr{L}^2(\Omega)$-norm by H and in the $H_1(\Omega)$-norm by E. Let P be the orthogonal projection operator $\mathscr{L}^2(\Omega)\to H$. By means of a projection, equation (14) can be written as

$$u'+F(u)=Pf=g.$$

Suppose that the following conditions are fulfilled:

(*a*) If $m=2$, then $f\in M^2{}_{\text{comp}}(E^*)$; $\bar{f}$ is its uniform mean.

(*b*) If $m\leqslant 3$, then $f\in M^2{}_{\text{comp}}(\mathscr{L}^2(\Omega))$, $\bar{f}$ is its uniform mean.

It is known that the stationary equation $F(u)=\bar{g}$ has at least one solution $\mathring{u}\in E$.

(*c*) The operator $A = F'_u(\mathring{u})$ ($A: E \to E^*$ is strongly elliptic; see property 3°) does not have a spectrum on the imaginary axis.

From Theorem 2 with due regard for the important property 4° we obtain

Proposition 1. *For* $m = 2$ *equation* (14) *has a unique solution* u^ω *for which*

$$\sup_{t \in J} \|u^\omega - \mathring{u}\|_H + \sup_{t \in J} \int_0^1 \|u^\omega(t+s) - \mathring{u}\|_{\mathring{H}_1} \, ds \to 0.$$

This solution is conditionally exponentially stable in H.

The case $m = 3$ requires a modification of the energy method since property 4° does not hold for $\lambda = 1$ and $m = 3$. It is assumed that Ω is sufficiently smooth; then it is known that $\mathring{u} \in H_2(\Omega)$.

Proposition 2. *For* $m \leq 3$ *equation* (14) *has a unique solution* u^ω *for which*

$$\sup_{t \in J} \|u^\omega - \mathring{u}\|_{\mathring{H}_1(\Omega)} + \sup_{t \in J} \int_0^1 \|u(t+s) - \mathring{u}\|_{H_2(\Omega)} \, ds \to 0.$$

This solution is conditionally exponentially stable in E.

The proof of Proposition 2 imitates the main arguments of the energy method, but instead of a triple E, H, E^* we must take the triple H_2, E, H. Let us note that for the equation $u' + Au = g$ we have the estimate

$$\|u(t)\|_E^2 + \int_{t_0}^{t_0+1} (\|u\|_{H_2(\Omega)}^2 + \|u'\|_H^2) \, ds$$
$$\leq c_1^* \left\{ \|u(0)\|_E^2 + \int_{t_0}^{t_0+1} \|g\|_H^2 \, ds \right\} \quad (t_0 \leq t \leq t_0 + 1),$$

which is completely analogous to the estimates (1) and (2) (see Ladyzhenskaya [68], Russian p. 127). By proceeding as in Theorem 1, from condition (*c*) we find that $L = d/dt + A$ has an inverse $L^{-1}: X \to V$, where the spaces X and V differ from those introduced earlier by replacing the triple E, H, E^* by $H_2(\Omega), E, H$. It remains to show that the operator F acts from V into X and is continuously differentiable. Since the basic classes of functions are preserved under the projection operator P, it suffices to solve this problem for the operator $\frac{1}{2}\partial(u)^2/\partial x = uu_x = \Phi(u)$.

We set $Q=[0,1]\times\Omega$ and $\|u\|_p=(\int_\Omega |u|^p\,dx)^{1/p}$. By Hölder's inequality and Sobolev's embedding theorem we have

$$\int_Q |vu_x|^2\,dx\,dt \leq \int_0^1 \|v\|_4^2 \|u_x\|_4^2\,dt,$$

$$\|u\|_4 \leq k\|u\|_{H_1(\Omega)}, \quad \|u_x\|_4 \leq k\|u\|_{H_2(\Omega)};$$

$$\int_Q |vu_x|^2\,dx\,dt \leq k_1 \sup_{0\leq s\leq 1} \|v\|^2_{H_1(\Omega)} \int_0^1 \|u\|^2_{H_2(\Omega)}\,dt,$$

$$\|vu_x\|_X \leq k_1\|v\|_V^2\|u\|_V^2. \tag{15}$$

It follows from (15) that Φ acts from V into X; it also follows that

$$\Phi'_u(v)h = v_x h + v h_x$$

and that $\Phi'_u(v)$ is continuous with respect to $u \in V$. For this it suffices to consider the relations

$$\Phi(u+h) - \Phi(u) - u_x h - h_x u = hh_x,$$

$$\Phi'_u(z)h - \Phi'_u(v)h = (z_x - v_x)h + (z-v)h_x,$$

and to apply an inequality of the form (15) to them. Here we can assume that Proposition 2 is proved.

The case of a zero mean ($\bar{f}=0$) is especially interesting. Here $\mathring{u}=0$ and $A=-P\Delta=-\tilde{\Delta}$, and consequently condition (c) holds automatically.

6 The problem on the whole space

The method developed in § 3 is directly applicable only to boundary-value problems in a finite domain (the requirement that the embedding $E\subset H$ is compact). A careful analysis shows that the problem in the whole space, in a reasonable formulation, can be studied by close methods. The 'reasonable formulation' requirement concerns above all the choice of the function spaces in which the operator L_ω is studied; these spaces must firstly be similar to the corresponding spaces for boundary-value problems (the spaces V, X, W), and secondly have a norm invariant under translations not only with respect to t but also x. In brief, for the whole space we need to establish the complete analogue of Theorem 1 (with precise estimates for L_ω that are uniform with respect to ω). However, this question will not be dealt with fully here. Instead we consider the more transparent averaging problem for second-order equations in the Hölder classes. We give the main attention to the new details that arise, which are generally typical for problems in the whole space. These details are connected with certain complications in the regularity problem.

We consider the uniformly parabolic operator $L = \mathrm{d}/\mathrm{d}t + A(t)$, where

$$-A(t)u = \frac{\partial}{\partial x_i}(a_{ij}(t, x)u_{x_j} + a_i(t, x)u) + b_i(t, x)u_{x_i} + a_0(t, x)u.$$

Let $D = J \times R^m$, $C = C(D)$ be the space of continuous functions $u(z) = u(t, x): D \to R^1$ with the sup norm, and $C^{\alpha,\alpha/2}(D)$ be the set of $u(t, x) \in C(D)$ for which the following norm is finite:

$$\|u\|_{C^{\alpha,\alpha/2}(D)} = \|u\|_C + \sup_{(t,x),(t,x')\in D} \frac{|u(t,x) - u(t,x')|}{|x - x'|^\alpha}$$
$$+ \sup_{(t,x),(t',x)\in D} \frac{|u(t,x) - u(t',x)|}{|t - t'|^{\alpha/2}}.$$

Suppose that the following conditions hold:

(1) $a_{ij}, b_i, a_i, a, \partial a_{ij}/\partial x, \partial b_i/\partial x \in C^{\gamma,\gamma/2}(D)$ $(\gamma \in (0, 1))$.

(2) The uniform means $\bar{a}_{ij}$, $\bar{b}_i$, $\bar{a}_i$, and $\bar{a}_0$ exist and are attained uniformly with respect to $x \in R^m$.

It then follows that the coefficients of the averaged operator have the same smoothness properties as in the unaveraged case, that is,

$$\bar{a}_{ij}, \bar{a}_i, \bar{b}_i, \partial \bar{a}_{ij}/\partial x, \partial \bar{b}_i/\partial x \in C^\gamma(R^m).$$

By analogy with the finite domain case, we introduce the space V consisting of functions $u(z)$ $(z \in D)$ with the norm

$$\|u\|_V = \left(\int_Q |u(z)|^2 + \left|\frac{\partial u}{\partial x}\right|^2 \mathrm{d}z\right)^{1/2},$$

where Q is the unit cube in D and the sup is taken over the whole unit cube.

By a solution of an equation of the form $Lu = f$ $(f \in C(D))$ we mean a $u(z) \in V$ satisfying the identity $\int_D u(z)L^*\phi(z)\,\mathrm{d}z = \int f(z)\phi(z)\,\mathrm{d}z$, where L^* is the formally adjoint operator and $\phi \in C_0^\infty(D)$. The result that $u(z)$ and the derivative $u_x(z)$ belong to the class $C = C(D)$ is obtained from the following important estimates:

$\|u\|_C \leq l\{\|u\|_V + \|f\|_C\}$ (The maximum principle; see Ladyzhenskaya, Solonnikov & Ural'tseva [67], Russian p. 225, English tr. p. 181)

$\|u\|_V \leq l\{\|u\|_C + \|f\|_C\}$ (The energy estimate [67], Russian p. 170, English tr. p. 139)

$\|u_x\|_{C^{\alpha,\alpha/2}} \leq l\{\|u\|_C + \|f\|_C\}$ (A Nash type estimate [67], Russian p. 246, English tr. p. 219)

We emphasise that these estimates remain uniform if we introduce a parameter ω in the operator $A(t)$.

Theorem 3. *Suppose that the averaged operator is regular, that is, $Lu = f$ has a unique solution $u \in C$ for any $f \in C$. Then for $\omega \geq \omega_0$ the operators L_ω are uniformly regular with the estimate*

$$\|u\|_{C(D)} + \|u_x\|_{C^{\alpha,\alpha/2}(D)} \leq k\|L_\omega u\|_C \tag{16}$$
$$(\omega \geq \omega_0, \alpha \in (0, 1)).$$

Then if $\bar{f} = \bar{f}(x)$ is the uniform mean of $f(t, x)$, attained uniformly with respect to $x \in R^m$, and

$$L_\omega u^\omega = f(\omega t, x), \quad \bar{A}\mathring{u} = \bar{f},$$

then

$$\|u^\omega{}_x - \mathring{u}_x\|_C + \|u^\omega - \mathring{u}\|_C \to 0. \tag{17}$$

We preface the proof with several auxiliary propositions.

Proposition 3. *If L is weakly regular, then $L^*u = 0$ has no solutions in C.*

Proof. Let $L^*g = 0$ with $g \in C$. Since L is weakly regular, $Lu = g$ for some $u \in C$. Consider a smooth function $\phi(z)$ with compact support such that $\phi(z) = 1$ for $|z| \leq 1$. Then $u_\varepsilon(z) = \phi(\varepsilon z)u$ satisfies $Lu_\varepsilon = \phi(\varepsilon z)g + \eta_\varepsilon(z)$. Since u, $u_x \in C(D)$, from the readily obtained explicit expression for η it follows that

$$|\eta_\varepsilon| \leq c_0\{\varepsilon|\phi'_z(\varepsilon z)| + \varepsilon^2|\phi''_{zz}(\varepsilon z)|\}.$$

Now by Green's formula we have

$$0 = \int_D (gLu_\varepsilon - u_\varepsilon L^*g)\,dz = \int_D g^2\phi(\varepsilon z)\,dz + \int_D g\eta_\varepsilon\,dz. \tag{18}$$

Because $\int_D |\eta|\,dz = o(1)$, by taking the limit as $\varepsilon \to 0$ we see that $\int_D g^2\,dz < \infty$. But since $\int_D |\eta_\varepsilon|^2\,dz = o(\varepsilon)$, the second term in (18) tends to zero as $\varepsilon \to 0$ (by the Cauchy–Bunyakovskii inequality). Hence we obtain $\int_D g^2\,dz = 0$, that is, $g \equiv 0$.

Proposition 4. *If L is correct and $L^*u = 0$ has no solutions in C, then L is regular.*

Proof. It is easy to show that the range $\mathscr{R}(L)$ of a correct operator L is not only closed but also locally closed in C. Therefore, it is enough to prove that $\mathcal{R}(L)$ contains the subspace $C_0(D)$ consisting of those $f \in C$ for which $f(z) \to 0$ as $z \to \infty$.

Suppose that $f_0 \in C_0$ but $f_0 \notin \mathcal{R}(L)$. Then we can find a functional $g \in C^*(D)$ orthogonal to a subspace of $\mathcal{R}(L)$ and such that $(g, f_0) \neq 0$. The restriction of g to $C_0(D)$ defines a non-trivial finite measure μ. Therefore, we have $\int_D L\phi \, d\mu = 0$ $(\phi \in C_0^\infty(D))$. This means that the measure μ is a generalised solution of $L^*u = 0$. From the hypoellipticity properties of a parabolic operator (see Eidel'man [118], pp. 213–14, English tr., pp. 218–20), it follows that $L^*u = 0$ has a non-trivial sufficiently smooth solution $u(z)$ belonging to $\mathcal{L}^1(D)$. This solution lies in $C(D)$ since

$$\sup_{Q(1)} |u(z)| \leq c \int_{Q(2)} |u(z)| \, dz,$$

where $Q(1)$ and $Q(2)$ are arbitrary concentric cubes with sides of length 1 and 2, respectively (see, for example, Eidel'man [118], pp. 215–16, English tr. pp. 221–23). This proves Proposition 4.

Now we introduce all possible operators $\bar{L}_h$ whose coefficients are obtained from $\bar{L}$ by a translation followed by a local passage to the limit with respect to $x \in R^m$.

Proposition 5. *The operators $\bar{L}_h$ are weakly regular.*

Proof. Let $\bar{L}_h$ be obtained from $\bar{L}$ by a limiting translation with respect to $\{x_k\} \subset R^m$. Let $f \in C(D)$, $f_k = f(t, x - x_k)$, and $\bar{L}u_k = f_k$. We put $v_k = u_k(t, x - x_k)$. It follows from Nash's estimate that $\{v_k\}$ is locally compact in $C(D)$. By taking the limit we obtain that $\bar{L}_h v = f$, $v \in C(D)$ (as yet we have not established that the solution is unique). Proposition 5 is proved.

Proposition 6. *The operators $\bar{L}_h{}^*$ are correct.*

Proof. For simplicity of notation we assume that we are speaking of the operator $\bar{L}^*$. Assuming that it is not correct we obtain sequences $\{u_n\}$, $\{f_n\} \subset C(D)$ such that $\bar{L}^* u_n = f_n$, $\|u_n\|_C = 1$ and $\|f_n\|_C \to 0$. We select a $z_n \in D$ such that $|u(z_n)| \geq \frac{1}{2}$ and put $v_n = u_n(z + z_n)$. Since $\{v_n\}$ is locally compact in $C(D)$, by taking the limit we obtain a non-trivial solution of an equation of the form $\bar{L}_h{}^* u = 0$, which contradicts Propositions 5 and 3.

From Propositions 3–6 it follows immediately that the operators $\bar{L}_h$ and $\bar{L}_h{}^*$ are regular; in other words, the conditions of Theorem 3 are in fact symmetric in L and L^*.

Now we briefly outline a proof of Theorem 3. First we must show that the L_ω $(\omega \geqslant \omega_0)$ are uniformly correct. Assuming otherwise we have a sequence $\omega_n \to \infty$ and sequences $\{u_n\}, \{f_n\} \subset C$ such that $L_\omega u_n = f_n$, $\|u_n\|_C = 1$, $\|f_n\|_C \to 0$. We take a point z_n such that $|u_n(z_n)| = u_n(x_n, t_n) \geqslant \frac{1}{2}$ and put $v_n(z) = u_n(z + z_n)$. It is important that $\{v_n\}$ is locally compact in C (a Nash type estimate); we may assume that $v_n \to_{\text{loc}} \mathring{v}$. Now we must realise a passage to the limit similar to that in Theorem 1. A small feature is that as well as the main averaging procedure, in this case we have a passage to the limit with respect to x. Therefore, $\mathring{v}$ is a solution not necessarily of $\bar{L}u = 0$ but of some equation $\bar{L}_h u = 0$.[4] But this contradicts the regularity of $\bar{L}$.

The uniform correctness of $L_\omega{}^*$ $(\omega > \omega_0)$ is proved in exactly the same way. But then the regularity of L_ω follows directly from Proposition 4. The proof of the remaining assertions of the theorem should not present difficulty.

In conclusion we mention a most commonly used sufficient condition for the regularity of $\bar{L}$: $\bar{a}_0(x) \leqslant 0$ and every $\bar{a}_{0h}(x) \not\equiv 0$. This condition is fulfilled if, for example, $\bar{a}_0 \leqslant 0$, $\bar{a}_0 \in \mathring{C}(R^m)$ and $\bar{a}_0 \not\equiv 0$.

Comments and references to the literature

Bogolyubov's lemma is proved in his book [8]. A very extensive bibliography on the averaging method is given in the monograph of Mitropol'skii [90]. The contents of Chapter 11 are mainly from Zhikov's article [56]. The results in § 6 were obtained by Zhikov jointly with L. Tsend and M. Otel'baev (unpublished). A technically different approach to averaging in parabolic problems which has been developed by Simonenko [101] must be mentioned.

[4] At this point it is important that the mean values of the coefficients are uniform with respect to $x \in R^m$.

Bibliography

1 Amerio, L. (1955). Soluzioni quasi-periodiche, o limitate, di sistemi differenzali non lineari quasi-periodiche, o limitati, *Annali di Matematica Pura ed Applicata* (4), **39**, 97–119.

2 Amerio, L. & Prouse, G. (1971). *Almost periodic functions and functional equations*, New York and London, van Nostrand–Reinhold.

3 Arnol'd, V. I. (1975). *Matematicheskie metody klassicheskoi mekhaniki* (Mathematical methods of classical mechanics), Moscow, 'Nauka'.

4 Baskakov, A. G. (1970). On the almost-periodic functions of Levitan, in *Studencheskie raboty*. Voronezhskii Gosudarstvennyi Universitet, pp. 91–4.

5 Baskakov, A. G. (1973). Criteria for almost-periodicity, *Trudy Matematicheskogo Fakul'teta Voronezhskii Gosudarstvennyi Universitet*, **8**, 1–8.

6 Bogolyubov, N. N. (1939). Some arithmetic properties of almost periods, *Zapiski Kafedry Matematichno Fiziki Institutu Bud Vel'ko Mekhaniki, Akademiya Nauk Ukrainskoi SSR*, **4**.

7 Bogolyubov, N. N. (1948). An application of the theory of positive definite functions, *Sbornik Trudov Instituta Matematiki, Akademiya Nauk Ukrainskoi SSR*, **11**, 113.

8 Bogolyubov, N. N. (1945). *O nekotorykh statisticheskikh metodakh v matematicheskoi fizike* (On some statistical methods in mathematical physics), Kiev, Akademiya Nauk Ukrainskoi SSR.

9 Bogolyubov, N. N. & Krylov, N. M. (1934). *Novye metody nelineinoi mekhaniki* (New methods in non-linear mechanics), Kiev, 'Naukova Dumka', pp. 54–84.

10 Boles Basit, R. (1971). Connection between the almost-periodic functions of Levitan and almost automorphic functions, *Vestnik Moskovskogo Universiteta, Seriya* 1. *Matematika i Mekhanika*, **26** (4), 11–15.

11 Boles Basit, R. (1971). A generalization of two theorems of M. I. Kadets on indefinite integrals of almost-periodic functions, *Matematischeskie Zametki*, **9**, 311–21 (*Mathematical Notes*, **9**, 181–6).

12 Boles Basit, R. (1971). Some problems of the theory of abstract almost periodic functions, Ph.D. dissertation, Moscow State University.

13 Boles Basit, R. & Zhikov, V. V. (1971). Almost-periodic solutions of integro-differential equations in a Banach space, *Vestnik Moskovskogo Universiteta, Seriya* 1. *Matematika i Mekhanika*, **26** (1), 29–33.

14 Boles Basit, R. & Tsend, L. (1972). A generalized Bohr-Neugebauer theorem, *Differentsial'nye Uravneniya*, **8**, 1343–8 (*Differential Equations*, **8**, 1031–5).
15 Bohl, P. (1893). Uber die Darstellung von Funktionen einer Variablen durch trigonometrische Reihen mit mehreren einer Variablen proportionalen Argumenten, Magister dissertation, Dorpat.
16 Bohl, P. (1906). Uber eine Differentialgleichung der Storungstheorie, *Crelles Journal*, **131**, 268–321.
17 Bohr, H. (1925). Zur Theorie der fastperiodischen Funktionen, I, *Acta Mathematica*, **45**, 29–127.
18 Bohr, H. (1925). Zur Theorie der fastperiodischen Funktionen, II, sActa Mathematica, **46**, 101–214.
19 Bohr, H. (1934). Again on the Kronecker theorem, *Journal of the London Mathematical Society*, **9**, 33–6.
20 Bohr, H. (1930); (1935). Kleinere Beitrage zur Theorie der fastperiodischen Funktionen, *Danske Videnskabernes Selskab Matematisk-Fysiske Meddeleser*, **10**, 8.
21 Bohr, H. (1932). Uber fastperiodische ebene Bewegungen, *Commentarii Mathematica Helvetici*, **4**, 51–64.
22 Bohr, H. (1934). *Fastperiodische Funktionen*, Berlin, Springer-Verlag. (Translation (1934): *Pochti-periodicheskie funktsii*, Moscow, 0G1Z.)
23 Bohr, H. & Neugebauer, O. (1926). Uber lineare Differentialgleichungen mit konstanten Koeffizienten und fastperiodischer rechter Seite, *Nachrichten von der Gesellschaft der Wissenschaften zu Göttingen, Mathematisch-Physikalische Klasse*, 8–22.
24 Bochner, S. (1959). *Lectures on Fourier integrals*, Annals of Mathematics Studies, No. 42, Princeton, N.J., Princeton University Press. (Translation (1962): *Lektsii ob integralakh Fur'e*, Moscow, Fizmatgiz.)
25 Bochner, S. (1927). Beiträge zur Theorie der fastperiodischen Funktionen, I, *Mathematische Annalen*, **96**, 119–47.
26 Bochner, S. (1933). Fastperiodische Lösungen der Wellengleichung, *Acta Mathematica*, **62**, 227–37.
27 Bochner, S. (1933). Abstrakte fastperiodische Funktionen, *Acta Mathematica*, **61**, 149–84.
28 Bochner, S. (1962). A new approach to almost periodicity, *Proceedings of the National Academy of Sciences of the USA*, **48**, 195–205.
29 Bochner, S. & von Neumann, J. (1935). On compact solutions of operational differential equations, *Annals of Mathematics* (2), **36**, 255–91.
30 Brodskii, M. S. & Mil'man, D. P. (1948), On the centre of a convex set, *Doklady Akademii Nauk SSSR*, **59**, 837–40.
31 Bronshtein, I. U. (1975). *Rasshireniya minimal'nykh grupp preobrazovanii* (Extensions of minimal groups of transformations), Kishinev, 'Shtinitsa'.
32 Bronshtein, I. U. & Chernyi, B. F. O. (1974). Extensions of dynamical systems with uniformly asymptotically stable points. *Differentsial'nye Uravneniya*, **10**, 1225–30 (*Differential Equations* **10**, 946–50).
33 Wiener, N. (1930). Generalized harmonic analysis, *Acta Mathematica*, **55**, 117–258.
34 Veech, W. A. (1965). Almost automorphic functions on groups, *American Journal of Mathematics*, **87**, 719–51.
35 Wolf, F. (1938). Approximation by trigonometrical polynomials and almost periodicity, *Proceedings of the London Mathematical Society*, **11**, 100–14.

36 Gel'fand, I. M. (1938). Abstrakte Funktionen und lineare Operatoren, *Matematicheskii Sbornik* 4, 235–86.
37 Gorin, E. A. (1970). A function-algebra variant of a theorem of Bohr–van Kampen, *Matematicheskii Sbornik*, **82**, 260–72 (*Mathematics of the USSR – Sbornik*, **11**, 233–43).
38 Gottschalk, W. A. & Hedlund, G. A. (1955). *Topological dynamics*, Providence, R. I., American Mathematical Society.
39 Daletskii, Yu. L. & Krein, M. G. (1970). *Ustoichivost' reshenii differentsial'nykh uravnenii v Banakhom prostranstve*, Moscow, 'Nauka'. Translation (1974): *Stability of solutions of differential equations in a Banach space*, Providence, R. I., American Mathematical Society.
40 Dunford, N. & Schwartz, J. T. (1958), *Linear operators. Part I: General theory*, New York – London, Interscience. (Translation (1962): *Lineinye operatory. Obshchaya teoriya*, Moscow, Inostr. Lit.)
41 Demidovich, B. P. (1967). *Lektsii po matematicheskoi teorii ustoichivosti* (Lectures on the mathematical theory of stability), Moscow, 'Nauka'.
42 Doss, R. (1961). On bounded functions with almost periodic differences, *Proceedings of the American Mathematical Society*, **12**, 488–9.
43 Zhikov, V. V. (1965). Abstract equations with almost periodic coefficients, *Doklady Akademii Nauk SSSR*, **165**, 555–8 (*Soviet Mathematics Doklady*, **6**, 949–52).
44 Khikov, V. V. (1966). On the harmonic analysis of bounded solutions of operator equations, *Doklady Akademii Nauk SSSR*, **169**, 1254–7 (*Soviet Mathematics Doklady*, **7**, 1070–3).
45 Zhikov, V. V. (1967). Almost periodic solutions of differential equations in a Banach space, *Teoriya Funktsii, Funktsional'nyi Analiz i ikh Prilozheniya*, **4**, 176–87.
46 Zhikov, V. V. (1970). Almost periodic solutions of linear and non-linear equations in a Banach space, *Doklady Akademii Nauk SSSR*, **11**, 278–81 (*Soviet Mathematics Doklady*, **11**, 1457–61).
47 Zhikov, V. V. (1969). A problem of Bochner and von Neumann, *Matematicheskie Zametki*, **3**, 529–38. (*Mathematical Notes*, **3**, 337–42).
48 Zhikov, V. V. (1970). A supplement to the classical Favard theory, *Matematicheskie Zametki*, **7**, 239–46. (*Mathematical Notes*, **7**, 142–6).
49 Zhikov, V. V. (1971). The existence of solutions almost periodic in the sense of Levitan for linear systems (second supplement to the classical Favard theory), Matematicheskie Zametki, **9**, 409–14 (*Mathematical Notes*, **9**, 235–8).
50 Zhikov, V. V. (1969). The problem of almost periodicity for differential and operator equations, *Sbornik Nauchnye Trudy, Vladimirskii Vechernii Politekhicheskii Institut*, **8**, 94–188.
51 Zhikov, V. V. (1971). Some remarks on the compactness conditions in connection with a paper of M. I. Kadets on the integration of abstract almost periodic functions, *Funktsional'nyi Analiz i ego Prilozheniya*, **5**, 30–6. (*Functional Analysis and its Applications*, **5**, 26–30).
52 Zhikov, V. V. (1973). Monotonicity in the theory of almost periodic solutions of non-linear operator equations, *Matematicheskii Sbornik*, **90**, 214–28 (*Mathematics of the USSR – Sbornik*, **19**, 209–23).
53 Zhikov, V. V. (1975). Some new results in abstract Favard theory, *Mathematicheskie Zametki*, **17**, 33–40 (*Mathematical Notes*, **17**, 20–4).

54 Zhikov, V. V. (1975). The solvability of linear equations in the Bohr and Besicovitch classes of almost periodic functions, *Matematicheskie Zametki*, **18**, 553–60 (*Mathematical Notes*, **18**, 918–22).
55 Zhikov, V. V. (1972). On the theory of the admissibility of pairs of function spaces, *Doklady Akademii Nauk SSSR*, **205**, 1281–3 (*Soviet Mathematics Doklady*, **13**, 1108–11).
56 Zhikov, V. V. (1973). The averaging principle for parabolic equations with variable principal term, *Doklady Akademii Nauk SSSR*, **208**, 32–5 (*Soviet Mathematics Doklady*, **14**, 26–30).
57 Zhikov, V. V. & Levitan, B. M. (1977). Favard Theory, *Uspekhi Mathematicheskikh Nauk*, **32** (2), 123–71 (*Russian Mathematical Surveys*, **32** (2), 129–80).
58 Zhikov, V. V. & Tyurin, V. M. (1976). The invertibility of the operator $d/dt+A(t)$ in the space of bounded functions, *Matematicheskie Zametki*, **19**, 99–104 (*Mathematical Notes*, **19**, 58–61).
59 Jessen, B. (1935). Uber die Sakulerkonstanten einer fastperiodischen Funktion, *Mathematische Annalen*, **111**, 355–63.
60 Jessen, B. & Tornehave. (1945). Mean motions and zeros of almost periodic functions, *Acta Mathematica*, **77**, 137–279.
61 Yosida, K. (1965). *Functional analysis*, Berlin and New York, Springer-Verlag. (Translation (1967): *Funktsional'yni analiz*, Moscow, 'Mir').
62 Kadets, M. I. (1958). On weak and strong convergence, *Doklady Akademii Nauk SSSR*, **122**, 13–16.
63 Kadets, M. I. (1968). The method of equivalent norms in the theory of abstract almost periodic functions, *Studia Mathematica*, **31**, 34–8.
64 Kadets, M. I. (1969). The integration of almost periodic functions with values in a Banach space, *Funktsional'nyi Analiz i ego Prilozheniya*, **3**, 71–4 (*Functional Analysis and its Applications*, **3**, 228–30).
65 Corduneanu, C. (1968). *Almost periodic functions*, New York, Interscience.
66 Krasnosel'skii, M. A., Burd, V. S. & Kolesov, Yu. S. (1970). *Nelineinye pochti-periodichieskie kolebaniya*, Moscow, 'Nauka'. Translation (1973): *Nonlinear almost periodic oscillations*, New York, Wiley.
67 Ladyzhenskaya, O. A. Solonnikov, V. A. & Ural'tseva, N. N. (1967). *Lineinye i kvazilineinye uravneniya parabolicheskogo tipa*, Moscow, 'Nauka'. Translation (1968): *Linear and quasilinear equations of parabolic type*, Providence, R.I., American Mathematical Society.
68 Ladyzhenskaya, O. A. (1970). *Matematicheskie voprosy dvizheniya vyazkoi neszhimaemoi zhidkosti* (2nd revised augmented edition), Moscow, 'Nauka'. Translation (1969): *The mathematical theory of viscous incompressible flow* (1st edition revised), New York, Gordon and Breach.
69 Lax, P. D. & Phillips, R. S. (1967). *Scattering theory*, New York, Academic Press.
70 Levin, B. Ya. (1948). A new construction of the theory of the almost periodic functions of Levitan, *Doklady Akademii Nauk SSSR*, **62**, 585–8.
71 Levin, B. Ya. (1949). On the almost periodic functions of Levitan, *Ukrainskii Matematicheskii Zhurnal*, **1**, 49–100.
72 Levin, B. Ya. & Levitan, B. M. (1939). On the Fourier series of generalized almost periodic functions, *Doklady Akademii Nauk SSSR*, **22**, 543–7.

73 Levitan, B. M. (1938). A new generalization of the almost periodic functions of H. Bohr, *Zapiski Mekhaniko-Matematicheskogo Fakulteta Khar'kovskogo Matematicheskogo Obshchestva*, **15**, 3–32.
74 Levitan, B. M. (1947). Some questions in the theory of almost periodic functions II, *Uspekhi Matematicheskikh Nauk*, **11** (6), 174–214.
75 Levitan, B. M. (1937). On an integral equation with almost periodic solutions, *Bulletin of the American Mathematical Society*, **43**, 677–9.
76 Levitan, B. M. (1953). *Pochti-periodicheskie funktsii* (Almost periodic functions), Moscow, Gos. Izdat. Tekhn-Teor. Lit.
77 Levitan, B. M. (1966). Integration of almost periodic functions with values in a Banach space, *Izvestiya Akademii Nauk SSSR, Seriya Matematika*, **30**, 1101–10.
78 Levitan, B. M. (1967). On the theorem of the argument for an almost periodic function, *Matematicheskie Zametki*, **1**, 35–44 (*Mathematical Notes*, **1**, 23–8).
79 Lions, J.-L. (1969). *Quelques méthodes de résolution des problèmes aux limites nonlinéaires*, Paris, Dunod-Gauthier Villars. (Translation (1972): *Nekotorye metody resheniya nelineinykh kraevykh zadach*, Moscow, 'Mir').
80 Lyubarskii, M. G. (1972). An extension of Favard theory to the case of a system of linear differential equations with unbounded Levitan almost periodic coefficients, *Doklady Akademii Nauk SSSR*, **206**, 808–10 (*Soviet Mathematics Doklady*, **13**, 1316–19).
81 Lyubich, Yu. I. (1960). Almost periodic functions in the spectral analysis of operators, *Doklady Akademii Nauk SSSR*, **132**, 518–20 (*Soviet Mathematics Doklady*, **1**, 593–5).
82 Loomis, L. H. (1960). Spectral characteristics of almost periodic functions, *Annals of Mathematics* (2), **72**, 362–8.
83 Lyusternik, L. A. (1936). Basic concepts of functional analysis, *Uspekhi Matematicheskikh Nauk*, **1**, 77–140.
84 Maizel', A. D. (1954). On stability of solutions of systems of differential equations, *Trudy Ural'skogo Politekhnicheskogo Instituta*, **51**, 20–50.
85 Marchenko, V. A. (1950). Methods of summation of generalized Fourier series, *Zapiski Nauchno-Issledovatel'skogo Instituta Matematiki i Mekhaniki i Khar'kovskogo Matematicheskogo Obshchestva*, **20**, 3–32.
86 Marchenko, V. A. (1950). Generalized almost periodic functions, *Doklady Akademii Nauk SSSR*, **74**, 893–5.
87 Massera, J. L. & Schäffer, J. J. (1966). *Linear differential equations and function spaces*, New York, Academic Press. (Translation: *Lineinye differentsial'nye uravneniya i funktsional'nye prostranstva*, Moscow, Izdat. 'Mir').
88 Millionshchikov, V. M. (1965). Recurrent and almost periodic trajectories of non-autonomous systems of differential equations, *Doklady Akademii Nauk SSSR*, **161**, 43–5 (*Soviet Mathematics Doklady*, **7**, 534–8).
89 Millionshchikov, V. M. (1968). Recurrent and almost periodic limit solutions of non-autonomous systems, *Differentsial'nye Uravneniya*, **4**, 1555–9 (*Differential Equations*, **4**, 799–801).
90 Mitropol'skii, Yu. A. (1971). *Printsip usredneniya v nelineinoi mekhanike* (The averaging method in non-linear mechanics), Kiev, 'Naukova Dumka'.
91 Mishnaevskii, P. A. (1971). An approach to the almost periodic regime and the almost periodicity of solutions of differential equations in a

Banach space, *Vestnik Moskovskogo Universiteta. Seriya Matematika i Mekhanika*, **3**, 69–76.

92 Montgomery, D. & Samelson, H. (1943). Groups transitive on the n-dimensional torus, *Bulletin of the American Mathematical Society*, **49**, 455–6.

93 Muckenhoupt, C. F. (1929). Almost periodic functions and vibrating systems, *Journal of Mathematical Physics*, **8**.

94 Mukhamadiev, E. (1971). The invertibility of differential operators in the space of functions that are continuous and bounded on the real axis, *Doklady Akademii Nauk SSSR*, **196**, 47–9 (*Soviet Mathematics Doklady*, **12**, 49–52).

95 Nemytskii, V. V. & Stepanov, V. V. (1949). *Kachestvennaya teoriya differentsial'nykh uravnenii*, Moscow, Gos. Tekh-Teor. Lit. Translation (1960): *Qualitative theory of differential equations*, Princeton, N.J., Princeton University Press.

96 Pełczynski, A. (1957). On B-spaces containing subspaces isomorphic to the spaces c_0, *Bulletin de l'Académie Polonaise des Sciences Class* III, **5**, 797–8.

97 Perov, A. I. & Ta Kuang Khai (1972). The almost periodic solutions of homogeneous differential equations, *Differentsial'nye Uravneniya*, **8**, 453–8 (*Differential Equations*, **8**, 341–5).

98 Perron, O. (1930). Die Stabilitatsfrage bei Differentialgleichungen, *Mathematische Zeitschrift*, **32**.

99 Pontryagin, L. S. (1954). *Nepreryvnye gruppy*, Moscow, Gostekhizdat. Translation (1966): *Topological groups*, New York, Gordon and Breach.

100 Reich, A. (1970). Präkompakte Gruppen und Fastperiodizität, *Mathematische Zeitschrift*, **116**, 216–34.

101 Simonenko, M. B. (1970). Justification of the averaging method for an abstract parabolic equation, *Doklady Akademii Nauk SSSR*, **191**, 33–4 (*Soviet Mathematics Doklady*, **11**, 323–5).

102 Sobolev, S. L. (1945). Sur la présque périodicité des solutions de l'équation des ondes I, II, III, *Doklady Akademii Nauk SSSR*, **48**, 542–5; **48**, 618–20; **49**, 12–15.

103 Sobolev, S. L. (1950). *Nekotorye primeneniya funktsional'nogo analiza v matematicheskoi fizike*, Leningrad, Leningrad Gos. Universitet. Translation (1963): *Applications of functional analysis in mathematical physics*, Providence, R. I., American Mathematical Society.

104 Stepanov, V. V. (1926). Uber einige Verallgemeinerungen der fastperiodischen Funktionen, *Mathematische Annalen*, **95**, 437–98.

105 Wallace, A. D. (1955). The structure of topological semigroups, *Bulletin of the American Mathematical Society*, **61**, 95–112.

106 Favard, J. (1927). Sur les équations différentielles à coefficients présque-périodiques, *Acta Mathematica*, **51**, 31–81.

107 Favard, J. (1933). *Leçons sur les fonctions présque-périodiques*, Paris, Gauthier-Villars.

108 Furstenberg, H. (1961). Strict ergodicity and transformations of the torus, *American Journal of Mathematics*, **83**, 573–601.

109 Furstenberg, H. (1963). The structure of distal flows, *American Journal of Mathematics*, **85**, 477–515.

110 Flor, P. (1967). Rythmische Abbildungen abelscher Gruppen II, *Zeitschrift für Wahrscheinlichkeitstheorie und Verwandte Gebiete*, **7**, 17–28.

111 Følner, F. (1954). Generalization of a theorem of Bogolyubov on topological abelian groups, *Mathematica Scandinavica*, **2**, 5–19.
112 Foias, C. & Zaidman, S. (1963). Almost periodic solutions of parabolic systems, *Annali della Scuola Normale Superiore di Pisa*, **3**, 247–62.
113 Cheresiz, V. M. (1972). Uniformly *V*-monotonic systems. Almost periodic solutions, *Sibirskii Matematicheskii Zhurnal*, **13**, 1107–22. (*Siberian Mathematics Journal*, **13**, 767–77).
114 Shcherbakov, B. A. (1966). Recurrent solutions of differential equations, *Doklady Akademii Nauk SSSR*, **167**, 1004–7 (*Soviet Mathematics Doklady*, **7**, 534–8).
115 Shcherbakov, B. A. (1972). *Topologicheskaya dinamika i ustoichivost' po Puassonu reschenii differentsial'nykh uravnenii* (Topological dynamics and the Poisson stability of solutions of differential equations), Kishinev, 'Shtinitsa'.
116 Shcherbakov, B. A. (1973). A general property of compact transformations of abstract functions, *Izvestiya Vysshikh Uchebnykh Zavedenii, Seriya Matematika*, **11**, 88–96.
117 Hardy, G. H., Littlewood, J. E. & Polya, G. (1952). *Inequalities* (2nd edition), Cambridge University Press. (Translation (1948): *Neravenstva*, Moscow, Inostr. Lit.).
118 Eidel'man, S. D. (1964). *Parabolicheskie sistemy*, Moscow, 'Nauka'. (Translation (1969): *Parabolic systems*, Amsterdam, North-Holland; Groningen, Wolters-Noordhoff).
119 Ellis, R. (1958). Distal transformation groups, *Pacific Journal of Mathematics*, **8**, 401–5.
120 Esclangon, E. (1904). Les fonctions quasi-périodiques, Thesis, Paris.
121 Esclangon, E. (1919). Nouvelles recherches sur les fonctions quasi-périodiques, Annales de l'Observatoire de Bordeaux.
122 Esclangon, E. (1915). Sur les integrales bornées d'une équation differentielle linéaire, *Comptes Rendus Hebdomadaires de Séances de l'Academie des Sciences*, Paris, **160**, 475–8.

Additional references

Sell, G. R. (1973). Almost periodic solutions of linear partial differential equations, *Journal of Mathematical Analysis and Applications*, **42**, 302–12.

Fink, A. M. (1974). *Almost periodic differential equations*, Lecture Notes in Mathematics, vol. 377, Berlin–New York, Springer–Verlag.

Brom, J. (1977). The theory of almost periodic functions in constructive mathematics, *Pacific Journal of Mathematics*, **70**, 67–81.

Shubin, M. A. (1978). Almost periodic functions and partial differential operators, *Uspekhi Matematicheskykh Nauk*, **32** (2), 3–47 (*Russian Mathematical Surveys*, **33** (2), 1–52).

Kozlov, S. M. (1978). Homogenization of differential operators with almost periodic rapidly oscillating coefficients, *Matematicheskii Sbornik*, **107**, 199–217 (*Mathematics of the USSR – Sbornik*, **35**, 481–98).

Zhikov, V. V. (1979). A pointwise stabilization criterion for second order parabolic equations with almost periodic coefficients, *Matematicheskii Sbornik*, **110**, 309–18 ((1980) *Mathematics of the USSR – Sbornik*, **38**, 279–92).

Zhikov, V. V., Kozlov, S. M. & Oleinik, O. A. (1982). Homogenization of parabolic operators with almost periodic coefficients, *Matematicheskii Sbornik*, **117**, 69–85.
Kozlov, S. M., Oleinik, O. A. & Zhikov, V. V. (1981). Sur l'homogénéisation d'opérateurs différentiels paraboliques à coefficients presque périodiques, *Comptes Rendus des Séances de l'Académie de Sciences, Paris*, **293**, Series 1, no. 4, 245–8.
Zhikov, V. V. & Sirazhudinov, M. M. (1981). Homogenization of non-divergent second order elliptic and parabolic operators, and the stabilization of the solution of the Cauchy problem, *Matematicheskii Sbornik*, **116**, 166–86.

Index